WHALE IN THE DOOR

Caitlin Press Inc.
8100 Alderwood Road, Halfmoon Bay, BC V0N 1Y1
www.caitlin-press.com

Edited by Jane Silcott
Text and cover design by Vici Johnstone
Printed in Canada

Caitlin Press Inc. acknowledges financial support from the Government
of Canada and the Canada Council for the Arts, and the Province of
British Columbia through the British Columbia Arts Council and the
Book Publisher's Tax Credit.

Library and Archives Canada Cataloguing in Publication

Le Bel, Pauline, author
     Whale in the door : a community unites to protect BC's Howe
Sound / Pauline Le Bel.

ISBN 978-1-987915-48-8 (softcover)

     1. Howe Sound (B.C.)—Environmental conditions. 2. Howe
Sound
(B.C.)—Social life and customs. 3. Howe Sound (B.C.)—History. I.
Title.

FC3845.H73L4 2017   971.1'31   C2017-904594-6

Whale in the Door

A Community Unites to Protect BC's Howe Sound

Pauline Le Bel

Foreword by Elizabeth May, OC
Photos by Richard Duncan

Caitlin Press

Rich Duncan

*To the Skwxwú7mesh Úxwumixw/Squamish Nation,
whose members have welcomed me on their traditional territory,
shared their stories with me, and deepened my love and respect for
Atl'kitsem/Howe Sound.*

CONTENTS

ACKNOWLEDGMENTS

I write these words on unceded Coast Salish territory with full awareness that I am a visitor in Howe Sound, a guest in Atl'kitsem. I acknowledge my debt to the Skwxwú7mesh Úxwumixw/ Squamish Nation for the great care they and their ancestors have shown for the land all these thousands of years. I honour their deep connection to the land and the waters of Atl'kitsem, and their ways of seeing the world, which have helped to transform my own ways of seeing. I reciprocate by donating some of the proceeds of this book to Squamish Nation youth programs.

Whale in the Door was informed by a community of voices who shared their love, their frustrations, and their hopes and dreams for Atl'kitsem: the Squamish Nation warrior acting as guardian of the waters; the brilliant biologist and her team restoring the Squamish Estuary; the gregarious citizen scientist who keeps detailed data on spawning salmon and herring; the marine biologist seeking protection for forage fish habitat; the Squamish Nation Elder reviving the tradition of the seagoing canoe; the filmmakers capturing the beauty, the fragility and the stories of Howe Sound; the young Squamish Nation ethnobotanist re-establishing an important traditional food; the politicians promising jobs and economic activity; the Squamish Nation Chief attempting to balance prosperity for his people and protection of the land; the divers exploring the astonishing depths of the Sound; the self-named right-wing industrialist, providing safe homes for salmon fry under the Squamish Terminals docks; and the retired businesswoman who has spent the past several years of her life gathering these people together to discover the collective voice of Howe Sound.

Many thanks to my publisher, Vici Johnstone, and to all those who shared their stories and their expertise. Thanks to my friend Chris Corrigan, who read an early draft of chapter one and made an intriguing suggestion that sent me off on an unexpected direction; thanks to Julie Gardner, who was the first person to read *Whale* from beginning to end and offered many constructive suggestions based on her experience as a planner and consultant for First Nations in BC.

Thanks to biologist Edith Tobe, who introduced me to the Squamish Estuary and offered ongoing friendship and support; to Randall Lewis—the busiest man I know—who took the time to show me the impressive water restorations in Squamish territory, remains available for my many questions and continues to be a trusted ally; to John Buchanan for a splendid day in search of the Squamish River; and to Ruth Simons for introducing me to new ideas and new people and taking me on a boat tour of Howe Sound.

Many thanks to all those who read the manuscript from beginning to end and offered their comments, many of which I integrated: Julie Gardner, Ruth Simons, Edith Tobe, Mark Fettes, Chris Corrigan, Robert Ballantyne and editor Jane Silcott. Huge thanks to Paul for his understanding and support during the research, writing and rewriting of the book, and for dependable and cheerful chauffeur service.

Thanks to Diane Reid, who explained the difference between Glass Sponge Gardens and Glass Sponge Reefs and shared her photos of these magnificent Howe Sound creatures. And Rich Duncan, what a pleasure to ride in the Cessna 170 with you over the glaciers of Howe Sound; thanks for your dedication to capturing the beauty of Howe Sound with your camera, and for permission to include your photos in this book.

My deepest gratitude to the Squamish Nation: to Chief Ian Campbell, Chief Bill Williams, Roy "Bucky" Baker, Chris Lewis, Randall Lewis, Linda Williams, Joyce Williams, Leigh Joseph, Xwalacktun, Khelsilem, Bob Baker/S7aplek, Rebecca Duncan, Gloria Nahanee, Gwen Harry, Humteya, and the Squamish Valley

Bob Baker, who shared their stories, their culture and their language with me, and gave me permission to include their words in this book. Chen kwen mantumi!

Special thanks to Chris Lewis, Squamish Nation councillor, who gave me permission to include the legend given to him by his ancestors, a story that gave me the title for this book. Chris also helped me understand the importance of including these stories when he wrote: "Thanks for thinking of using our legend and my words to put it into a modern context."

By bringing the past into the present, it becomes clear that although there have been many attempts to erase the culture of the Squamish Nation, they are very much here.

As I travelled throughout Howe Sound, as I spoke with those who expressed their deep love and concern for the Sound, I sometimes felt I was having a conversation with Atl'kitsem herself, as if she were speaking to me. I have included what I heard and thank Atl'kitsem for adding her wisdom. Huy chexw a!

FOREWORD

— Elizabeth May

In the late 1980s, pollution of Howe Sound became a major environmental issue. Greenpeace made it national news by testing the sediments of Howe Sound for toxic contamination. The Greenpeace tests revealed alarming levels of dioxins and furans. Government scientists conducted their own tests and discovered that the levels of toxic chemicals were even higher than originally reported by Greenpeace, and that the local pulp mills at Port Mellon and Woodfibre had polluted the waters with industrial poison. Environment Canada made its test results public and began work to correct the contamination. But in addition to the toxins from the pulp mills, the heavy metal runoff from the local Britannia Mine had also added to the environmental destruction of Howe Sound. The crab fishery was closed due to pollution and new regulations were developed for pulp and paper effluent. In 1988, Woodfibre was told to clean up its act, and Port Mellon (now Howe Sound Pulp and Paper) was retrofitted into one of the cleanest pulp mills in the world.

In 2006, as Pauline Le Bel describes in this book, the federal and provincial governments finally responded by building a treatment plant for the mine's contaminated waters, and Woodfibre Pulp Mill shut down.

Fast forward to Squamish and Howe Sound today. Marine life is returning. The small forage fish, the herring and even the whales have made their return. Loss of the mill was not an economic death knell for the area. Without the belching mill, Howe

WHALE IN THE DOOR

Sound was transformed into a very attractive area for residential development. Ambitious private-sector investment backed the creation of one of Canada's most photographed tourist attractions. The Sea to Sky Gondola takes tourists up 2,900 feet above sea level. From the sparkling restaurant and viewing areas, you have a breathtaking view of the now uncontaminated Howe Sound. There is a suspension bridge to appreciate the life of the canopy forest. And right smack in the middle of that awesome view, a giant LNG plant, ironically named after the Woodfibre pulp mill, is now planned for this delicate area.

In a further irony, of the nineteen LNG projects proposed by the former BC Liberals, the first to clear federal-provincial environmental assessment was the Woodfibre LNG facility at the head of Howe Sound. The approval of the project by the federal Minister of Environment and Climate Change is an example of the short-sighted regulatory system left in the wake of previous Prime Minister Stephen Harper's re-alignment of environmental laws. Due to the spring 2012 omnibus budget bill, C-38, protection of our fisheries and the rigour of environmental assessment were slashed. Unlike the Environment Canada of the 1980s, the new Environment Canada 2.0 is focused on assisting proponents in gaining approvals.

No doubt, significant improvements have occurred due to the agreement between the developer and the Squamish First Nation in changing the planned water-cooling system. Because of its incredible volume, the chemically treated heated water does damage, particularly in smaller enclosed waterways, like Howe Sound or the Saanich Inlet. Thanks to the ongoing process of Squamish First Nation environmental review, the Indonesian developer of the LNG plant has conceded that the plan for once-through cooling will be abandoned. Instead, the plant will employ the superior, but more expensive, closed-loop-cooling system.

But what is most disturbing is that the proposal for the new Howe Sound Woodfibre plant received federal approval before anyone could determine if LNG tankers can safely transit the area. Unlike crude oil or bitumen, LNG is highly volatile and flammable. Taking a gas and making it a liquid allows it to be greatly reduced in volume, but once liquefied it behaves differently than when it is a gas. While an LNG accident would not coat our coastlines with a toxic mess, a pierced hull could result in the LNG pooling above ocean water. As the LNG returns to its gaseous state, its volume could expand to six hundred times that of its liquid state, creating a highly dangerous vapour cloud floating over the ocean, or depending on the wind, over local communities. Unlike the US, Canada has no regulations in place for required shipping zones to protect shoreline communities.

But the economics of LNG globally have also changed in recent years. In 2014, total global shipments of LNG were 243 million tonnes per annum (MTPA). By 2021, the total capacity expected from all nations, but not including Canada, is an amount almost equal to current shipments at 240.45 MTPA.

The previous BC government committed to producing almost the same volume of LNG as the total current global supply, but global demand for LNG is only increasing 2 percent a year. There is a really large, logical problem here. By the time LNG plants are even under construction in BC, the global LNG supply will exceed demand. Huge investments in Australia, where natural gas is produced from old coal seams, have already proven to be a bust as governments increasingly bail out a losing investment.

Pauline Le Bel's *Whale in the Door* tells the story of resource development in BC and explores its impact on local communities, Indigenous cultures and our precious oceans. She is a writer who lives on Bowen Island, and a soon-to-be neighbour of the Woodfibre LNG plant. Unless...

—Elizabeth May, OC

> Member of Parliament Saanich - Gulf Islands
> Leader of the Green Party of Canada

An Invitation from Atl'kitsem/ Howe Sound

Welcome all you who long to explore, to be reconnected to the wild. I welcome you into my home. Pull on your gumboots, your hiking boots. Go barefoot if you dare, and paddle my waters, wander my rocky paths, climb my precipitous peaks. You won't need wi-fi here. I offer you a deeper connection.

Let the sounds of my wildness guide you. The silver song of Shannon Falls, the primeval cry of the great blue heron, the siren call of the Squamish Wind. You may not understand everything you see and hear. That's to be expected. You have to stay in one place long enough before the spirits of a place will speak to you, before you begin to comprehend the dreams of the wild.

You do not need to understand everything. Keep your mind and heart open, and dance respectfully and enthusiastically with everything you encounter. Listen to the stories told by my volcanoes, my forests, my rivers, my people—the Rooted People, the Winged People, the Swimming People, the Six-Legged, the Four-Legged and the Two-Legged.

In this place, you may come to understand the meaning of your own life. Allow yourself to be surprised, to be amazed by the symmetry that is my rainforests, the rhythm that is my tides, the majesty that is my mountains. Allow yourself to weep for the frayed edges, the mighty stumps in my forests, the creeks that no longer flow to the sea. Allow your tender heart to break. And remember to return here again and again. To witness.

THE CALL

We're on a bus speeding along the Sea to Sky Highway—one of the most picturesque roads on the planet. This bountiful arm of land and water at the eastern edge of the Salish Sea offers non-stop views of snow-capped wilderness. On our right, the Coast Mountains—custodians of lush coniferous forests and billow-ing waterfalls—rise straight up from the road. To the left, a wild archipelago of tree-clad islands and the blue waters of Howe Sound. Picture a triangular-shaped body of water, the base of the triangle meeting the Strait of Georgia between West Vancou-ver and the Sunshine Coast, and extending forty-two kilometres northeast to its crown at Squamish.

On the early morning ferry, we sailed away from the womb-like harbour we call Snug Cove, leaving behind Bowen Island/Kwilákm, our home in the genteel part of Howe Sound. After a brief but dramatic ocean cruise on the Queen of Capilano, we arrived at the ferry terminal in Horseshoe Bay/Ch'axa'y. A former sleepy coastal town, today it is the transportation hub for over two-and-a-half million vehicles and seven million passen-gers who take ferries to Bowen Island, the Sunshine Coast and Vancouver Island.

We're on our way in a chartered bus to reach the wilder, windier Howe Sound. Our destination: the town of Squamish, ancestral and modern-day home of Skwxwú7mesh Úxwumixw, the Squamish Nation, the First Peoples of Howe Sound. We've been summoned by Joyce Williams of the Squamish Nation. "God forbid our land should be totally devastated," she posted on the Skwomesh Action Group website, "and the children ask: What did you do to stop it?"

We're travelling on unceded Coast Salish territory. The historical evidence shows that these people have lived here since the last great Ice Age. The longevity of the Squamish Nation and their culture dates back to a time before the pyramids. My companions are twenty-four Bowen Islanders. They tell me they want to keep the "natural" in Super, Natural BC. They're missing most of the spectacular scenery because of the pouring rain, but their spirits are not dampened.

They've been fired up by industrial salesmen who promote a cooling, liquefying and exporting terminal in Squamish for the liquefied natural gas industry. The company's pervasive commercials paint a pretty picture of pipelines in the estuary and super tankers, burdened with methane, cruising our narrow inlet.

None of these Bowen Islanders are buying the BC premier's story that "liquefied natural gas is how we will meet the central challenge of our generation." They disagree on what, in fact, is the central challenge. They only know they need to think about the next generation, and that means protecting the waters of Howe Sound.

The highway, completed in 1959, is an impressive engineering achievement. Glacial valleys are U-shaped with flat bottoms and very steep sides. The highway is above sea level, and much of the route is either cut into the rock slope or built on bridges cantilevered out from the slope. The bus sways as we navigate ascents and descents, twists and turns. Signs warn of possible rock falls, and caution against stopping along sections of the highway. Other signs promise impressive viewpoints. Nobody asks to get out in the rain. The signage here is bilingual: English and Squamish, out of respect for the First Peoples, and an acknowledgement of the evolving and respectful relationship between the two cultures.

I'm covering this event for the *Undercurrent*, the Bowen Island weekly newspaper. Because islanders tend to live in a state of blissful insularity and focus on island matters, I'm curious as to why they are taking this trip. As the bus careens around another bend, I stumble from seat to seat to find out why people are

THE CALL

willing to get wet and cold, why something happening a ferry ride and forty kilometres away should be of concern to them. They carry signs saying, "Water is more precious than oil," "No LNG, No Pipelines, No Tankers" and "Save Our Sound." They range in age from eleven to eighty-three, from newcomers to old-timers, most of them unaccustomed to the activist role. They tell me they're on the road because they feel called to stand up for what they love.

The plan is to meet up in Squamish with five hundred other residents of Howe Sound and march to the LNG office. They tell me the proposed Woodfibre liquified natural gas plant is a threat to the future, that it will pose an unacceptable health and safety risk for all communities in Howe Sound. They believe pipelines burrowing through the Squamish Estuary and supertankers travelling the Sound will harm marine and terrestrial ecosystems. They tell me there are only seventy-eight southern resident orcas in the Pacific Northwest and that they must be protected from the noise of tanker traffic. They also voice concern about the many other development proposals coming forward for the Sound, including a gravel mine on McNab Creek, an important salmon-bearing estuary; a road through the Squamish Estuary; and a multi-million-dollar real estate development.

My friend Chris Corrigan offers another compelling reason for being on the bus: "This is a local and important act of reconciliation and support for those who host us on this land. We're being invited to come and be helpful by showing up. The important message of reconciliation is that we can do more together than separately."

Atl'kitsem/Howe Sound has known violence, intrusion. She is prepared for the unexpected, the strange, the surprising. She remembers, almost fondly, the crucible of the glacier and the great transition that occurred there. After an extended icy gestation, she birthed a unique geological entity. Midwifed by the spirits of the land throughout her labour, she felt the waters break and flow, and as the glacier moved off, she sighed in relief. Exhausted and spent, she thrilled at the sight of her magnificent

17

mountains, crowning. The ordeal was over, As the fjord was delivered, she cried out with joy.

Even now, you can still hear the birth cries as they continue to echo from mountain peak to mountain peak, moving the tides, pulling the rain down from the skies. Nothing has been forgotten of the wild ways. The memory shimmers in every leaf of every magnificent cedar tree, in every volcanic pebble, every drop of rain.

Today Atl'kitsem/Howe Sound is in the midst of another birthing process. This time the crucible is the mindset of a human culture that chooses growth over continuity, speed over grace and short-term financial gain over long-term well-being. The midwives attending this birth are those who listen to the land, who stand for the forests, who restore the waters, who document the progress of the homecoming salmon in a ritual millions of years old, who appreciate how humans are entangled in the ancient and ever-new web of being.

They have learned to see with new eyes. Eyes that really see the sun rising above the Coast Mountains, the wild salmon leaping up precipitous waterfalls to complete their destiny. Eyes that see source, and not just resource, that see the whole kit and caboodle, every nook and cranny, every drop of water, every herring egg as alive and kicking, connected to everything else, and asking for time to heal. Asking us to step back, listen and reflect.

Towering over us, unseen but felt, a pair of graceful peaks shimmers above the North Shore Mountains. An iconic landmark, visible from almost everywhere in the greater Vancouver area, they were named the Lions in the late nineteenth century. The village of Lions Bay, the Lions Gate Bridge and the BC Lions football team were named in their honour.

⋞⊚

The way you name something, a person, a tree or a mountain, is how you start to tell a story about it, a story that has a deep connection to that place or person. The name becomes a container for the stories, the things that happened. The Skwxwú7mesh people call those peaks Chíchiyuy, the Twins or the Sisters, in honour of two young women who brought peace to the land and the people.

～

I've heard many members of the Squamish Nation tell the wonderful story of Chíchiyuy (pronounced Chee-chee-oye) and I've read several versions in books. About two years ago at the Vancouver Public Library, I heard storyteller Rebecca Duncan tell an animated version. She made the story come to life and helped the listener see how this ancient story was still very relevant today. Rebecca gave me permission to tell a shortened version of it here.

In those days, the Squamish were often at war with the Northern People who would steal their women and bring them back as slaves. A man, a Squamish siyam, a noble person, had twin daughters who were as beautiful as they were hard working. Everyone in the village admired their skill at cleaning deer hides, making clothing from cedar bark and drying salmon.

One day while out berry picking, they were captured on a raid by the Northern People. These young women had been properly brought up. They had been taught to remain dignified and respectful no matter the circumstance, and that is how they came to be revered by the Northern People. They took such good care of their captors that they were greatly admired by them, and so it was no surprise when twin brothers, good hunters and fishermen, asked to be married to them. The sisters agreed to the marriage if their father would consent. A scout was sent out to seek the father's permission.

Their father said yes, but only if they would come to his village to celebrate the marriage. The northerners arrived with their moose meat and their fine blankets made of goat hair, and the father set a splendid table with salmon, eulachon oil, seaweed, and roots and berries in intricately designed cedar baskets. There was a great reunion of the sisters and their family, and much feasting and drumming and dancing. Many gifts were given.

The power of this celebration uniting two families in marriage was so great, it led to a treaty to end the wars that had gone on for many years. In honour of this peaceful coming together, the Sky Brothers transformed the twin sisters into those beautiful peaks you see today. The siyam's daughters have stood in this high place for thousands of years. Chíchiyuy will continue to stand for thousands of years to come, guarding the land, the sea and the mountains, reminding all who gaze upon them that there are peaceful ways to resolve differences.

We reach Porteau Cove, a protected marine park with a provincial campground, a mecca for scuba divers. Known to the Squamish as Xwaxw'chayay, a name referring to the sturgeon traditionally caught there, it is rich in marine and other wildlife and is one of the oldest archaeological sites on the coast, dating back ninety-eight hundred years.

Soon, the shorelines close in. We are held in a rocky landscape softened by the many waterfalls flowing from the mountains. Howe Sound, a narrow finger of ocean, displays her fjordal self. The bus drives past Britannia Beach, a once-thriving community with a checkered history of mining, as well as devastating avalanches and floods. The BC Museum of Mining, a tourist attraction, perches on the property of the former Britannia Copper Mine, a major polluter of Howe Sound. Britannia Beach has experienced the best and the worst of the BC boom-and-bust economy. Many Howe Sounders are unwilling to repeat the experience with fracked gas, pipelines and supertankers.

On our right, Sky Pilot Mountain stands tall over Britannia Beach. This jutting profusion of ice and rock is a popular destination for advanced backcountry skiers, who are attracted by its domes, ridges and glades where snow lingers well into spring. The mountain is also famous for a beer named in its honour. Howe Sound Brewing Company, located in Squamish, created the Sky Pilot Pale Ale; but there's nothing pale about the mountain, its jagged peaks reaching through the clouds, a challenge for experienced climbers. Sky Pilot feeds into the picture-postcard Shannon

Falls, which we are approaching. I make a mental note to take the Sea to Sky Gondola—when it's not raining—to savour the spectacular views.

We arrive at the head of Howe Sound, the town of Squamish in the Squamish Valley. "Hard-wired for Adventure" is the current slogan the district uses to promote itself. LNG is probably not the adventure they had in mind. The word Squamish is a loose adaptation of Skwxwú7mesh, meaning "Mother of the Wind." And the Squamish wind, which howls through the Sound, has earned an international reputation with windsurfers in search of thrills. The area is also a recreational paradise for heli-skiers, hikers, ice and rock climbers, cross-country skiers, mountain bikers, sailors and kayakers.

We step off the bus, our nostrils filling with moist, salty air. We are now completely embraced by the Coast Mountains and the fascinating geological stories they tell of plate tectonics and the collision of continental plates. As I walk along the beach and take in the hundreds of people who are arriving to protest, I imagine the collisions in the coming months and years in the environmental and political arenas. We stand in silence, our heads tilted to admire the beauty at the summit of our inlet.

We underestimate beauty, take it for granted. But beauty moves people in a way that data and statistics can't. Beauty makes people care. We take in the breathtaking views of Shannon Falls —among the highest in the world—welcoming almost half a million people every year, the Tantalus Mountain Range to the west, and to the south, the Stawamus Chief, a vast pillar of rock born one hundred million years ago in the belly of the Earth, sculpted by flowing glacial ice.

I take a really good look at the Chief, beloved by climbers from all over the world. Towering more than two thousand feet above the highway, it's a daunting presence, the second-largest granite monolith in the world. Gibraltar is the largest. Every year, hundreds of thousands of visitors hike the routes to the three peaks at the back of the Chief, and climb the mammoth rock face.

But do they know the story? People see the mountain. It's beautiful. Impressive. They love to climb it. I've heard they travel from all parts of the world to stand in front of that magnificent mountain to get married! But do they know the story? You don't just climb a mountain. You look for the spirit. It's right there in the story. The Skwxwú7mesh call this mountain Siyam Smanit. It does not mean the Chief. Smanit means mountain. Siyam means someone who is deeply respected. A leader, a teacher, an Elder, not a chief. This mountain has great spiritual significance for the First People. And don't point. It's a sign of disrespect.

There are many stories about Siyam Smanit. Here's one I have often heard. Thank you to Rebecca Duncan for approving this telling.

There was a time very long ago when four brothers, supernatural beings called Xaays, came to this land to give teachings. They were Transformers who used their powers to bring balance to the world. They changed people and animals into the mountains and rocks you see today. In those days, Siyam Smanit was a longhouse. And at that time, there was a big potlatch going on, with all kinds of animals and people present. Xaays transformed the longhouse into a mountain, with all the spirits stuck inside. They're still there to this day. The mountain has become a symbol of togetherness. Maybe that's why people are drawn to it when they take their marriage vows.

Howe Sound is a great teacher, a teacher of humility. The size and extent of the formidable mountains, the forceful wind, the powerful (and at times frightening) ocean currents all inspire deep respect and grand stories. To the northeast is Mount Garibaldi, British Columbia's most famous volcano. The mountain was named after an Italian patriot. The Squamish know it as Nch'kay, pronounced Ch'i Kai—Mountain of Dirty Snow—for the snowfields that are covered with dust from the lava beds.

There is a saying among the Squamish that "Mount Garibaldi always covers its face," and today is no exception.

A thick blanket of cloud hides the impressive peak of rock and ice, which began erupting about 250 thousand years ago, when the Squamish Valley was filled with a large river of ice that flowed into what is now Howe Sound. The western flank of the mountain was built on top of this glacier. A magnificent peak, almost nine thousand feet above sea level, it has been dormant in modern times but is still considered potentially active. Garibaldi/ Nch'kay is part of the volcanic belt known as the Pacific Ring of Fire, stretching from Japan, north to Alaska and south to South America.

When I was in my fifties, I climbed Black Tusk in Garibaldi Provincial Park with friends. It was a hot day in late August and the sun was high and bright as we worked our way along the dusty switchbacks, relieved somewhat by wild blueberry bushes offering the occasional burst of refreshing purple juice. By the end of the hike, I was sweaty, exhausted, covered in dust, ready for a bracing and very brief dip in the glacial turquoise waters of Garibaldi Lake. Growing up in the city had made me far too dependent on speed, convenience and comfort. It never occurred to me that the mountains might contain stories, let alone spirit.

I imagine how the early inhabitants might have felt when they first looked up at the Coast Mountains. Small and insignificant perhaps. Mountains give you the right size for yourself. The spirit of the place is formidable, the need to tell stories strong. According to Squamish Nation oral history, the First Peoples anchored their canoes to Mount Garibaldi during the Great Flood, and waited for the waters to recede. Today, we the Second Peoples are also looking for a place to anchor our collective canoe as the flood of re-industrialization threatens our home.

There are lots of smiles and greetings as the community gathers, many of them meeting for the first time around this common concern. They are dressed in every imaginable form of rain gear as they juggle umbrellas and signs. The mood is worried, angry and celebratory, all at the same time. We know this wild beauty is worthy of every effort to protect it. We are also aware that it must be protected from ourselves, and that our human ingenuity, greed and industriousness must be respectful of the ways and needs of this very special place, so it may thrive and the lands and waters may continue to support us.

I meet Joyce Williams of the Skwomesh Action Group, who organized the march. Joyce is a strong, beautiful woman. Her eyes moisten as she sees the number of people walking toward her. "Thank you for coming," she says and takes my hand. I tell her it was her love for her home—our home—and her powerful words that brought us here.

Joyce tells me the march is a way for her to show her family she will take action to protect the land. Joyce's mother, Linda Williams, admits she was concerned no one would come to stand with them in the pouring rain. Watching the long line of people gather to march through the town of Squamish, she says: "It was like your first love, your heart beating fast." As we march to the Woodfibre office on the main street, I sense many hearts beating as one.

"We will always be here," says Linda. "We're not going anywhere. This land and the waters is who we are, it's where we come from and we'll always fight for it." Her words are echoed by the marchers' footsteps. This is their home too; they are also part of the Howe Sound story. They want to have a say in the writing of the next chapter. Governments, they've been told, may grant permits but only communities can grant permission. And so far, no one has asked for permission.

"Everything is to be understood in terms of the way it is interwoven with the rest of the universe," wrote twentieth century philosopher Alfred North Whitehead. He was challenging the basic assumptions of our modern industrial culture, where everything is isolated, independent, disconnected from everything else. The marchers are here to reimagine Howe Sound. They carry a vision of an interconnected fjord made up of communities of people, animals, plants and fish, interwoven with the land, air, waters. They are asking the question, "What kind of world do we want to live in?"

When you listen to the waters, the wind, the forests, you begin to see how the land takes care of things in its own good time. The land is meant to shape you, shape how you live, not the other way around. The deer, the salmon, the whales, all the animals, all the water creatures know this. The First People know this. They know the berries, the salmon, the cedar, the sun's energy are all part of a big beautiful weaving. And this weaving is held together by the ancestors' bones. Holy ground.

When I asked anthropologist and author Wade Davis what he thought of an LNG facility in Howe Sound, he said he was not against LNG in principal. "I worked for twelve years on the biggest LNG project in the world." Wade helped a Texas millionaire find the optimum location for his huge LNG project. However, he is against LNG in Howe Sound. "This is not simply about a local issue in Howe Sound," he told me. "This is a metaphor for who we are as a people. If we are actually prepared to invest our lives in this way, the most glorious fjord in the world, what else in our country will be immune to such violations? Putting an LNG facility in Howe Sound would be like logging Stanley Park."

In his book *The Master and His Emissary: The Divided Brain and the Making of the Western World*, Iain McGilchrist insists on the need for myth and metaphor in order to understand the world. "Such myths or metaphors are not dispensable luxuries, or 'optional extras,' still less the means of obfuscation: they are fundamental and essential to the process. We are not given the option not to choose one, and the myth we choose is important: in the absence of anything better, we revert to the metaphor or myth of the machine."

We arrive at the Woodfibre LNG office in downtown Squamish singing the Women's Warrior Chant, an energizing Indigenous song. The rain has not let up, so we watch the proceedings through a blaze of colourful umbrellas. As the rain blows sideways with the force of the wind, we listen to a welcoming song, drumming, heartfelt speeches, more drumming, and an outpouring of love for this beautiful, vulnerable inlet.

At the end of the proceedings, the bus arrives and twenty-five soggy, cold and grateful people jump aboard and are whisked down the highway and back to Bowen Island. As the rain plays the bus like a steel drum, I reflect on this place I call home. How the mountains, the rainforests, even the rain, are becoming guides and friends, wisdom keepers. I begin to write my column for the newspaper, the letters blending into each other on the wet paper. I know this is just the beginning of a much longer journey.

Margaret Wheatley has written extensively about community. She believes "there is no power for change greater than a community discovering what it cares about." If she's right, we could see the proposed re-industrialization of Howe Sound as an awkward gift. One that compels our community to discover what it cares about and guides us to weave a realistic and wholesome dream for our unique fjord.

Although I've lived in Howe Sound for nineteen years, I've experienced it in small ways. Most of them near my home on Bowen Island. It's a geographical thing. To go anywhere in Howe Sound from any other part involves a ferry ride, sometimes two, and a car. I make a promise to myself to venture away from my island sanctuary and become an explorer, to travel through the communities of Howe Sound, and become more intimately acquainted with the mountains, rivers and forests, as well as the diverse human and other-than-human populations.

I will stop to look at what's above, at what I have yet to notice. I will look at what's below, flipping over rocks, examining what's underneath, searching, sniffing, walking this way and that, a bent stick in my hands, a water diviner, looking for the hidden, the underground, reeling with the stick's vibration upon each discovery. Like my companions on the bus and the marchers in Squamish, I will discover what I love about Howe Sound and be prepared to stand up for it. Atl'kitsem, this small, precious corner of the Earth, has claimed me. There is no turning back. I will be in service to this place, become the eyes and ears of this place, become entangled in her stories.

THE GENIUS OF ATL'KITSEM/HOWE SOUND

In his advice to the gardeners and architects of his day, the eighteenth-century poet Alexander Pope told them to "consult the genius of a place in all, that tells the waters, or to rise, or to fall. All must be adapted to the genius of the place," he wrote, and nothing be "forced into it, but resulting from it." For this genius "paints as you plant, and as you work, designs." Even a brief visit to Howe Sound allows one to observe a genius that has evolved over millennia, and continues to design and to paint in the sea, rivers, mountains and forests.

For most of its dramatic life, Howe Sound was a river valley. Thousands of years ago, the valley became covered in ice. Only the highest mountain peaks poked above the vast icefields. At the end of the Ice Age, an immense glacier of slow-moving ice streamed southward, gouging and scraping and shaping the valleys around the Squamish and Cheakamus rivers, then pushing through the river valley that is now Howe Sound. As the glacier melted and receded, it carried sand, gravel and meltwater and designed a long, deep, narrow valley. When the sea levels began to rise, a unique geological character was created: a fjord with a series of small inlets and islands.

As part of coastal BC, the Sound is at the leading edge of the westward-moving North American tectonic plate. It overrides the oceanic plate, moving at the languorous but steady speed of four centimetres per year. The geological history of Howe Sound is an ongoing saga of rifts, subductions, tectonic pressures, eruptions and collisions. These collisions design mountains, earthquakes, volcanoes and, it appears, passionate people.

Howe Sound paints with water: the glaciers, the rivers, the lakes, the waterfalls, the brackish waters of the estuaries, the rhythm of the ocean tides ebbing and flowing, moving over six hundred million tonnes of water a day, the rain filling the rivers, flooding the valleys, the rain, the rain, the rain relentlessly returning to the sea. The Squamish River is the soul of Howe Sound, the source of 90 percent of its fresh water.

"Rivers are the arteries of our planet," says Mark Angelo, conservationist and creator of World Rivers Day. "Lifelines in the truest sense." The Squamish River, the main artery of Howe Sound, designs with sediment. The river carries the sediment and fresh water, rich in nutrients, to the mouth of the sea, creating an estuary conditioned by marine tides, waves and salty water. According to scientists, this sediment is filling in Howe Sound by one metre per year—geology on steroids, as one scientist described it.

We are surrounded by the genius of organisms that have lived gracefully in the mountains and the waters for millennia. One of the most fascinating designs of Howe Sound is also the most ancient: glass sponge reefs, or bioherms, are a rare and astonishing underwater formation, the oldest and most exquisite marine feature on the planet. Fossil records indicate that over the Earth's history, glass sponge reefs have created the largest biological structures that have ever existed on the planet, with some encompassing nearly forty-three hundred square kilometres, almost double the size of the Great Barrier Reef.

These bioherms were common during the Jurassic period and thought to have become extinct sixty million years ago. Today they are found only on BC coasts, and there are many reefs available to scuba divers in Howe Sound. Their discovery just a few years ago was, according to paleobiologist Dr. Manfred Krautter, like discovering a living herd of dinosaurs.

Although they look like plants, sponges are in fact one of the oldest animals on Earth. Their delicate skeleton is made of carefully woven silica or glass, which they extract from seawater, creating an intricate structure of tubes and towers. Without lungs or mouths, they pump water through their bodies and filter oxygen and bacteria to breathe and feed. Just one sponge is capable of filtering nine thousand litres of water per day. They thrive on bacteria; glass sponges chew their way through 237 tonnes of it every day. They form a critical component of Howe Sound's marine environment, modifying currents, providing ideal habitat for multitudes of fish and other species, serving as refuge for juvenile rockfish, and filtering impurities in the ocean water.

Glen Dennison, a citizen scientist and former resident of Lions (or Sisters) Bay, has been observing and exploring the waters of the Sound for thirty years. He began cataloguing sponges using high-resolution mapping equipment to document the location and diversity of a dozen such reefs in Howe Sound. In his book *Diving Howe Sound Reefs and Islands*, he describes his first encounter with the sponge bioherms with all the excitement of an astronaut stepping on to the moon.

As he drifts through "an aquatic alien world…in search of the great Howe Sound bioherms, undiscovered for thousands of years," he dives down 125 feet to his undersea garden, "falling… dropping…you feel like Alice falling towards Wonderland…a kaleidoscope of glowing multi-coloured giant sponges with total coverage of the ocean floor.…These are the cold-water reef builders, our closest analogue to the tropical water corals. Hovering, we gently swim over the sponges, some standing six feet high. Light colours of yellow, orange and clean white…stems growing from stems…random forms that look more like clouds or something from a Dr. Seuss book. The sponges are beautiful beyond words, fragile as fine crystal, yet as large as a small car." Glen may have felt like Alice as he discovered the glass sponge reefs, but he was acting more like the prince who awakens Sleeping Beauty with his love.

Photographer Diane Reid, whose photos enhance Glen Dennison's book, has been diving since 1989 and is one of the regular group that dives with Glen. She documents their discoveries with her camera. "It's addictive," she tells me. "The glass sponges are breathtaking, all so unique, some shaped like vases, some tubular, others like elephant ears, and all the little creatures going in and out of them." Diane explained that in a sponge garden, the sponges grow on rock and are fairly common, while in the reefs or bioherms, the sponges grow on a base of dead sponges. It's the reefs that get scientists and divers really excited. "By sharing my pictures," she says, "I feel like I'm helping to protect Howe Sound."

Glen and the Marine Life Sanctuaries Society are instrumental in gaining protection for the bioherms, which can be damaged by crab and prawn traps as well as by trawling. According to the Canadian Parks and Wilderness Society, BC's massive reefs are also eminently suitable for World Heritage Status. They're that precious on a global scale. The glass sponges are a powerful reminder that there is so much we have to learn about our marine environment. We have a responsibility to proceed carefully with our desires to improve upon already whole and perfect nature.

The intricate web of life in the ocean depends on the marvelous plants. Without brains, hands or legs, they invented photosynthesis, which triggers phytoplankton to bloom, providing food for zooplankton such as krill, who feed small fish like herring, who in turn feed dolphins and whales. It's a marine food web that is an intimately connected design, one that is easily disrupted.

Salmon are also part of the inspired design of Atl'kitsem/ Howe Sound. They have been returning to the Sound for an estimated one million years in one of the planet's great migrations. They begin their life cycle in rivers and streams in the Sound. As youngsters, they head out for the open sea on an epic journey of up to a thousand miles, avoiding predators, garbage and toxic chemicals. They mature for several years, building up strength and muscle for the journey home to return to the same river, to scale the falls, leap for love and spawn on gravel beds. Salmon get one chance to mate. After spawning, they die and the cycle starts all over again. One can only imagine the wisdom and instinct, the genius that guides them.

There are several theories as to how they return to the exact place of their birth. One suggests that salmon can detect Earth's magnetic field and this helps them navigate over long distances in the ocean; once in the vicinity of the estuary or entrance to their birth river, they use their sense of smell to guide them the rest of the way home. Other theories suggest they navigate by the stars or the sun. However they do it, it's a remarkable feat that is part of the genius of this place. Five species of salmon live and die in Howe Sound: chinook, chum, coho, steelhead trout and numerous small pinks.

It's hard to think of any species that is more giving of itself than salmon, their silver fins leaping over death as they seek to spawn with a mate. The generosity of their numbers, and their long, arduous journey to return to their birthplace, represents perseverance and a selfless service to the cycle of life. Salmon are the life givers, offering food to more than two hundred species, including bears, wolves, otters, seals, orcas, eagles and humans. Their remains, deposited by bears, add rich nutrients, nitrogen, sulfur, carbon and phosphorus to the forests. When you eat salmon, you take in the wild energy of an ancient species that has the stamina, desire and endurance to travel thousands of miles to its birthplace, to ensure the survival of the next generation.

Salmon, according to Terry Glavin in *The Last Great Sea*, is the vital link between marine and terrestrial ecosystems, the foundation of the geology and the ecology of the west coast. "As salmon set about the long process of recolonizing the coastal landscape from their Ice Age refugia, the first trees to spring up from the glacial till of the valley bottoms took root in the spawned-out bodies of the first salmon…there isn't a forest on this coast that has not been home to salmon…there isn't a month of the year where salmon are not spawning somewhere on the coast." Salmon have shaped the west coast by simply being salmon.

꧁

Hold on a minute, you're missing the most important part of the story about Sts'úḵw'i7. That's what the Skwxwú7mesh people call them.

What you need to know about salmon is that they're not just another fish in the sea. The First People knew that. They knew Sts'úḵw'i7 were kin, humans who lived in longhouses in the sea. In the spring, they took on the silvery dress of the salmon and swam up the rivers to offer themselves as food when their Squamish brothers and sisters were hungry. After a long, hard winter, the Skwxwú7mesh were thrilled to see them. So many of them, they turned the rivers red, and you could walk across the river on their backs. You should have heard the great hullabaloo when that first Sts'úḵw'i7 arrived. It was celebrated with a grand welcome of ritual and song. As if a high-ranking chief had arrived. What a party!

Elders have told me how everyone in the village got to eat a bite of that first salmon, and not one morsel was wasted. Then they carefully returned Sts'úḵw'i7's bones to the sea so they would come back together. That way the salmon could revive, return to its home and regain its human form. When the rest of the salmon brothers and sisters arrived, some were sun-cured, some were dried in smokehouses over alder or hemlock fires, and stored for the winter. All the ceremony and feasting was done in accordance with agreements made between the Salmon People and the First People.

The Skwxwú7mesh Úxwumixw—Squamish Nation—belong to the linguistic and cultural groupings of the Coast Salish. They've lived and designed and painted in Howe Sound since time immemorial. Their teachings and practices come out of the land, out of a nine-thousand-year-old process of witnessing and sense-making. Their first villages were in the Squamish Valley, a lush valley for hunting, trapping and fishing at the mouth of the Squamish River. They later extended their territory throughout the inlet, and north and south to what we call Whistler, West and North Vancouver, False Creek and Burrard Inlet.

Around the year 1800, at least sixteen villages existed on the Squamish River. The homes were parallel to the shore, near the water, close to their canoes. In the old villages, large extended families lived in huge post-and-beam longhouses made of western red cedar. They were designed and constructed without nails. Each family had its own fire for cooking and keeping warm. They ate with stone knives and spoons made from the horns of mountain goats. The society was polygamous, and a chief often had four or five wives. They were happy and miserable like the rest of us. There were some bad apples in the bunch, and some really fine ones, just like the rest of the world.

They also had summer homes made of woven mats or slabs of cedar bark, supported by poles. A 1975 archaeological survey uncovered evidence of several summer camps or villages on islands in the Sound: nineteen sites on Gambier Island, whose spiritual name is St'a'pes; five sites on Keats Island; ten on Bowen/Kwilákm; seven on Pasley; and within the town of Gibsons/Sch'en'k, ten sites. Middens, granite hammers, a basal point adze and other tools and belongings showed where the First Peoples had lived and hunted and fished. There are about 180 recorded archaeological sites on Squamish territory, but the identification of all cultural and heritage sites that merit protection has only begun.

Most longhouses were used as community dwellings, while others were set aside for the exclusive use of winter spiritual dances. Felling the giant cedars for a longhouse required skill. First, a niche was carved into the trunk with hammer and stone wedge, then a slow-burning fire was started, while the tree trunk was kept wet. Eventually, the fire reached the other side of the trunk and the tree fell. This log was split into planks to form the walls of the longhouse.

The entrance to the longhouse was protected by large cedar house posts with the spirits of animals carved into them. Inside, people walked under a cedar bough to remove negative thoughts and feelings. The interior walls were lined with sleeping platforms covered with more cedar boughs; the mattresses were made of woven reed mats and animal skins. They knew how to live with the seasons. Winter, the rainiest, coldest season, was spent indoors doing ceremony, telling stories and eating the food that had been harvested and preserved in summer and fall.

At one point in their history, the Squamish Nation possessed over twenty longhouses; one of the larger longhouses was located at what is now called Lumberman's Arch in Stanley Park.

An important part of the ritual of the longhouse was—and continues to be—the witnessing ceremony, which could involve a memorial rite, a marriage, the initiation of a young person as a dancer, or the bestowing of a traditional name. Ceremonies took place in the winter when food reserves of dried salmon and berries had been stored. People congregated in their permanent villages and participated in a rich ceremonial life.

"Ceremonies were an integral part of First Nations people's daily lives," writes Lynda Gray, a member of the Tsimshian Nation who was raised in Vancouver, in her book *First Nations 101: Tons of Stuff You Need to Know about First Nations People.* "We had countless ceremonies and rituals serving distinct purposes that brought the community together in times of sorrow, prayer, transition and celebration. Some ceremonies were purely spiritual in nature, while others publicly marked milestones for individuals, families, or the community....Ceremonies passed from generation to generation, were given to us in dreams or ceremony, or were developed when the need arose."

In 2006, I was honoured to attend a witnessing ceremony on the Capilano/Xwemelch'stn Reserve in North Vancouver, at the site of one of the oldest Squamish villages. Xwemelch'stn translates into Fast Moving Water of Fish, and refers to the swift flowing Capilano River. The ceremony was a "bon voyage" for those who were setting out on the great Tribal Canoe Journey, an annual event celebrating the Coast Salish culture from the coast of Alaska to BC and Washington. It first began in 1989 as a way to revive the canoe culture, which had all but disappeared. Each year, a different nation hosts the journey, which can take up to a month.

The Xwemelch'stn longhouse was set up with tables, one for each canoe family; other tables were heaped with feasting dishes. Each family was invited to sing their canoe song, each one so different, so personal to their culture. Then the feasting began, and the Elders were asked to approach the tables and eat first. There was a moment of cultural queasiness on the part of my white-haired partner and I as we remained seated, unsure of the proper protocol, causing much waving and prodding on the part of our hosts to get up and join the Elders. It was explained to me later that, in their culture, the wisdom of Elders is respected. Elders are considered part of the genius of the place. They must eat first.

Midway through the proceedings, someone brought a toddler to the front of the room. The little girl had been wandering around unattended. The mother of the girl came forward, deeply humiliated, and retrieved her daughter. Losing your child is shameful. Later in the ceremony, there occurred something they called "the Work," where people are invited to speak their minds and hearts and be witnessed. A tall, powerfully built man came forward and in a soft, humble voice, confessed that he had been the one in charge of looking after his granddaughter, the one who had seemingly been abandoned. He had lost sight of her and as a result he had brought shame on his daughter. He apologized for the harm he had caused to both these people he loved.

I will never forget this father's public apology to his daughter, and the deep feelings of love, forgiveness and healing that were generated between father and daughter and everyone who opened their hearts to them. I experienced a profound appreciation for the ways of this ancient culture. This genius of ceremony, this power of passing on ways of knowing through ceremony, surely must have helped them to survive more than a century of depravations and hostility.

When I attend ceremonies, feasts and pow-wows, I can't help but notice the well-behaved children, and how they respect and listen to their Elders. In the old days, I was told, they used to scare the wee ones with stories of the Wild Woman of the Woods or Cannibal Woman. Ḵálḵalilh went out after dark to gather children who weren't in their beds. She picked them up, threw them in a bag, took them home, cooked them up and made herself a tasty meal. I suppose that's how they got the kids to bed early in those days. Today, Wild Woman of the Woods has been transformed into mosquitoes. She continues to enjoy children's blood.

Although they have been described as a peaceful people, the Squamish were sometimes at war with Lekwiltok and Haida villages, and held slaves as simple property, either taken or bought. We are fortunate to have recorded information about the Squamish way of life from the book *Conversations with Khatsahlano, 1932–1954*. The author, Major J.S. Matthews, was a Vancouver City archivist who recorded his conversations with August Jack Khatsahlano, a Chief of the Squamish Nation, about First Nations life in the Vancouver area. August Jack (1867–1971) was born in the village of Xway xway on the peninsula that is now called Stanley Park. His last name lives on, somewhat altered, in the community of Kitsilano, a name that offers great pleasure for the tongue as it travels around the front of one's mouth. You need to say it out loud to get the feel of it.

From Khatsahlano, we learn that the head of the longhouse, the siyam, was a man "whose wealth, generosity and general charisma earned him the respect of others." The siyam was "the man who says the most wise things." Siyams gave advice but that did not mean they had more control. Villages were generally governed by a male, and members of the village nobility who could be male or female. The Squamish people recognized and lived in alignment with the genius of Howe Sound and adapted to changes, ensuring that nothing be "forced into it but resulting from it."

The Squamish displayed their genius in an extensive use of plants for their everyday lives, according to ethnobiologist Nancy J. Turner, who has done major research with the Indigenous people of BC. In her book *Plant Technology of First Peoples in British Columbia*, she describes more than one hundred plants traditionally used by the Squamish and other First Nations for shelter, clothing, hats, nets, ropes, threads, tools, containers, decorations, scents, harpoon shafts, cleaning agents, paints, dyes, insect repellents and toys. All the domestic products required for the good life.

The Squamish had a sophisticated understanding and re-spect for the plant world and found uses for the many trees, bushes, kelp, grasses and roots in their environment. They dug the roots of plants with tough sticks or antlers and made teas with the dried roots. The roots of skunk cabbage or swamp lantern, prolific on the west coast, were collected from the ponds, roasted and peeled. The roots of bracken fern were peeled, roasted and made into cakes. They found food everywhere: in black moss—a lichen hanging from Douglas firs, fresh shoots from the salmon-berry bushes, and the northern rice root, whose roots look like a cluster of cooked rice. Dried and powdered, it became a staple of the winter diet, a good source of starch and sugar.

An extraordinary example of the genius of plant life in Howe Sound is the cedar tree—beautiful, symmetrical, growing to tremendous heights, soft and easy to split. The Coast Salish call it the Tree of Life, and the Squamish made the most of this blessing, using it to make skirts and ponchos, waterproof hats and baskets, tools for hunting and fishing, cooking boxes and giant dishes for feasting. Women worked with the soft inner bark of the cedar, pounding it until it was soft and fine enough to wrap a newborn babe. Shredded cedar bark also provided diapers and towelling. The men carved perfectly symmetrical canoes from the cedar—seagoing or cargo canoes, shovel-nose river canoes, and small canoes perfect for berry picking. For recreation, they raced in their canoes.

You need to remember that the First People lived in a world of spirit. All of this carving and pounding and shredding was accompanied by ceremony. Before you took the bark or cut down the cedar, you talked to the tree. Plants need to know why you are harvesting them. The tree has a spirit. You want the tree to know that it's taken for a good purpose. And while you work, you sing, you acknowledge your connection and your debt to the tree.

The Squamish have songs for everything, songs for greeting the morning, songs for working, for funerals, for attending children, songs for picking berries, songs for the change of the season, for making a canoe. Songs that get stirred up with drums, rattles made of deer hooves and the heartbeat of Mother Earth. Songs that come from the rhythms of the sea, the melodies of sweet birdsong, and the music of the winds playing in the trees.

Last year, I had the honour and pleasure of attending a canoe blessing at Ambleside. First there was the drumming and singing of traditional songs. Then the women cleansed the canoe with cedar boughs dipped in water to remove any negativity, so the canoe would travel lightly and safely in the water. Those who would paddle the canoe were also cleansed to release negative emotions so they could paddle as one in dangerous waters. I understood how the ceremony, and the singing and carving, were about transformation. For the Squamish, ceremony is what keeps everything alive. The heart, the mind, the land, the canoe, the waters.

The art of the Squamish Nation was present in their functional as well as their ceremonial objects. It was a vital part of their daily life. The spindle whorl, a traditional tool for spinning wool, is a good example. It had two parts: a slender shaft of wood about a metre long and a wooden disc with a central hole in which the shaft was inserted. The discs were intricately carved with images that were said to have a mesmerizing effect on the spinner as she twisted the fibres together to form one continuous thread. The images reflected the weaver's particular guardian spirit or that of the person who carved it. They were often carved with supernatural creatures who would lend spiritual energy to the wool, and in such a way that the spinning whorl would cause the image to change and move. The oldest known Salish spindle whorl dates back approximately twelve hundred years.

Wool for spinning came from mountain goats and small white woolly dogs. The whole process began with a trek up the mountains to gather the precious wool that was caught or tangled on low bushes. If the hunters came in with a full mountain goat, the women would pluck the fur, separating the long, coarse strands from the soft fur underneath. Then the fur was smothered in diatomaceous clay to draw out impurities and clean it.

Once the wool had been spun on the spindle whorl, it was woven on a loom formed from two vertical posts that supported two horizontal bars about six feet in length. The loom uprights were decorated with carvings and paintings. The weaver now had her sacred materials and could begin her work of transformation to create a ceremonial blanket. Colours were achieved from fern roots and lichen or the bark of the alder. For a nice yellow, Oregon grape roots were just the thing. The mordant to fix the dye was probably urine.

Girls as young as ten were trained by the grandmothers for this demanding art. Making a blanket was a spiritual experience that could take a very long time. Before being given the treasured wool, the weaver underwent ritual purification so she would put only positive energy into the blanket. As she wove, she put prayers and songs into the blanket, invoking the ancestors to ensure it would be strong. The eventual wearer of the blanket would be protected by those prayers.

Squamish blankets were not used for sleeping but for ceremonial purposes. They offered spiritual protection as well as protection from the elements. They were used for ceremonies, as potlatch gifts and as dowries. The blanket was an emblem of wealth, and the blanket weaver was highly respected for her ability to transform raw material into objects of great beauty, wealth and power. Blankets were highly valued and often used as currency for which other goods could be purchased or bartered.

Rebecca Duncan did not have the chance to observe the women weaving when she was a child. The resurgence began twenty years ago, she tells me, with women learning the techniques again. But she is on a mission to find a photo of her grandmother and her baskets. "There is only one photo and I'm going to find it." She wants to incorporate her grandmother's signature weave into her own weaving. Rebecca specializes in cedar capes. "My spirit soars when I weave with cedar," she tells me.

The eighteenth- and nineteenth-century journals of European explorers and traders report a well-established weaving tradition. When the Hudson's Bay came along with their blankets that you could buy for next to nothing, the art of weaving went into decline. Sixty fresh salmon, which were caught by the thousands, was all you needed to buy the Hudson's Bay variety. By the early twentieth century, blankets were rarely, if ever, made. Girls in residential schools, separated from their grandmothers and their training, were taught knitting and sewing, but not weaving. Women's work gravitated to the canneries.

I think about that woman weaving her beautiful ceremonial blanket, weaving her prayers into the blanket, working hard for many days to make a blanket that would be cherished by her people. Imagine the satisfaction she must have felt as her hands were guided by the ancestors, while her children played at her feet. How proud she must have felt that her blanket was a sign of prosperity and wealth. And how awful she must have felt when her work was replaced by the machine-made blankets. The new blankets meant a lot less work, but how could they protect the wearer when there were no prayers woven in, no sacred mountain goat wool, no sweat and spirit of the weaver?

The art of the Squamish Nation was shaped by the waters, the forests and the spirits who guarded and haunted them. As they worked with trees, they mastered techniques of carving and weaving cedar. "It is important to emphasize," writes Turner, "that the actual knowledge of weaving or carving techniques is a relatively small part of the entire knowledge needed: the times and locations for harvesting materials, ways to harvest without harming the plant populations, and the techniques for processing and storing the materials are just as critical to the production of the final products." Without the use of metal of any type, which was extremely rare, and pottery, which was virtually unheard of, plant materials were essential to survival.

Plants contain energy. Everything contains energy. All the things we see and all the things we don't see. The land breathes here. You can feel it. You can hear it. The mountains draw their breath from the wind, and this breath flows down into the valleys and enters everything, energizes everything. Every plant, every animal, every human. In June, I attended the National Aboriginal Day canoe races in Squamish. The wind was in charge of everything—the sea, the canoes, the trees, the hairdos. It was an invigorating afternoon.

I was surprised to learn that the Squamish did not make pottery. Many ancient civilizations used clay to build pots for cooking, for serving, for ceremonial purposes. One has to consult the genius of the place to find understanding. Part of the genius of Howe Sound is the energy contained in the rain, soil, climate and salmon, and how this energy supports the forests, especially the cedar. The ingenious use of cedar to make waterproof baskets and boxes for cooking and eating is the legacy of a people who have lived a long time with the land, working with the plants in a reciprocal way, and learning from trial and error and observation to trust the land deeply.

From Lee Maracle, a Coast Salish poet and author, I received a deeper appreciation of the sacred relationship between the Coast Salish and plants. "We would never have filled the soil around our villages with grass that needed constant cutting," she writes in her book *Memory Serves: Oratories*. "The cutting of a plant requires ceremony, for it is impairing, painful and insulting. We would never grow a plant that causes pain for the sake of aesthetics. Grass is intrusive now. It has replaced the medicines, the food and the animal life that grew here in abundance."

Lee's words prompted me to rethink my own gardening practices. To see my garden in a new way, a way that feels ancient, a way that honours the plants, and does the least to disturb them, a way that harvests with gratitude and tenderness. And yes, ceremony. Even something as simple as finding a way to sing my gratitude to the kale, the garlic, the peas, the apples.

The economy of the Squamish Nation was based on trade with other Salish nations. The Grease Trail, a historic overland trade route connecting the Pacific Coast with the interior, was a vital source for the economy of the day. For thousands of years the First Nations traders followed the trail, backpacking heavy baskets held in place by cedar ropes attached to headbands. The Squamish bartered with the Interior Salish, trading their highly valued eulachon oil/s7aynxw for furs and copper.

The Coast Salish, like most Indigenous peoples, had a non-western view of property. Ownership was not in the land, per se, but rather in access to salmon streams, herring spawning grounds, hunting grounds and berry patches. Occupancy was synonymous with Aboriginal ownership. Property also included names, stories, ceremonies and songs. To give someone the right to sing a tribal song was a precious gift.

☙

Don't forget to mention that the Skwxwú7mesh were a travelling people. They paddled up and down the Sound, along the rivers and inlets, and to the islands. Did you know there are two names for Howe Sound? One name for paddling up the Sound, Atl'kitsem—pronounced At-Kat-sum—and another name for paddling down, Texwnewets'—pronounced Chock-Now-it. Makes sense. It looks different whichever way you're travelling. In the summer, they migrated to summer villages for food. Ch'ekw'élhp, what Newcomers call Gibson's Landing, was plentiful with sea mammals. Xway xway and S7ens in Burrard Inlet were splendid places to harvest smelt.

One island was especially good for hunting ducks in the bay. There were plenty of mallards and surf scooters and mergansers. And sea lions in the reefs. Newcomers called this island Bowen Island, named for someone who never saw the sun rise or set on this beautiful place. One of the bays was loaded with clams, so the First People named it Kwilákm. It means Clam Bay. Used to be good eating on Kwilákm.

☙

There was no shortage of food. The land offered them elk, deer, bear, goat, as well as wild blueberry, blackberry, salmon berry, salal berry, grasses and roots, many roots from many plants. The sea provided them with mussels, crab, clams, seaweed, herring, trout, urchin, seal, eulachon and salmon. When the tide was out, the table was set.

Over many thousands of years, the Squamish continued to establish a gift relationship with the land, acknowledging their participation in, and dependence upon, the genius of Howe Sound. They had a profound belief in the power of place. Through cultural activities such as spirit dancing, the potlatch and winter ceremonies, they developed a sophisticated relationship with the land, one that allowed them to live in harmony for a very long time. This relationship, diminished by colonialism, is becoming whole again through the efforts of many to restore the land, the waters, the language and the food systems, as well as ceremony and purpose.

Howe Sound has its own unique genius. The rainforests, the rivers, the salmon, the cedar, the eagle, the mountains, the glass sponges are all teeming with intelligence and wisdom. This wisdom includes the ancestral knowledge of the Squamish Nation, a knowledge deeply rooted in place, earned over many generations of experience and observation, learning from the land, witnessing change and adapting to it. They have a complex cultural tradition that evolved over the centuries and continues to evolve. This is a wealth that must be protected. It will prove critical as we navigate the changes that are already occurring on the land and the waters.

The genius of Howe Sound speaks to us every day in the barking of sea lions, the *prraaawk prraaaw* of ravens, the comings and goings of the tides. It exists all around and within us, a genius that has perfected the art of resilience over thousands of years. The more we understand and appreciate the genius of Howe Sound, the more we will find our proper place in this story.

When I was writing *Becoming Intimate with the Earth* a few years ago, it became clear to me that we are not so much born into a place as we are born into the stories of that place. If I want to be truly at home in Atl'kitsem/Howe Sound, I need to immerse myself in the stories of this place. More important-ly, I need to understand that Atl'kitsem is the primary story, the ongoing drama. I am a stagehand, a bit player, opening my mind and heart to that intelligence, recognizing how it far sur-passes my own. A feeling of awe and reverence for this genius is the first step toward coming into alignment with the living, breathing, watery, wild, wise and extravagant fabric of life that is Atl'kitsem.

THE BETRAYAL

Eduardo Galeano, the Uruguayan writer, summed up the first contact between Indigenous and Europeans in this scathing parody of the colonialist perspective:

> In 1492, the natives discovered they were Indians,
> they discovered they lived in America,
> they discovered they were naked,
> they discovered there was Sin,
> they discovered they owed obedience to a King and a
> Queen from another world and a God from another heaven,
> and this God had invented guilt and clothing,
> and had ordered burned alive all who worshipped the Sun
> the Moon the Earth and the Rain that wets it.

According to written records, the first Europeans to arrive in Howe Sound were the Spanish in 1791. They named it Boca del Carmelo. One year later, it was Captain George Vancouver's turn. In his book *A Voyage of Discovery to the North Pacific Ocean and Round the World*, published in 1798, he described the islands he found in the Sound: "The shores of these, like the adjacent coast, are composed principally of rocks rising perpendicularly from an unfathomable sea; they are tolerably well covered with trees, chiefly of the pine tribe, though few are of luxuriant growth."

The pines he refers to were most likely Douglas fir, cedar and hemlock. For some reason that can only make sense to the eighteenth-century English mentality, Vancouver named the Sound after Admiral Earl Howe, rather than "the pines," "the perpendicular rocks" or "the unfathomable sea." Howe, a naval captain who had taken part in a British naval victory against some enemy or other, in some conflict or other, had never set eyes nor foot in the Sound. The naming of a place in this way is a daily reminder of colonial attempts to obliterate memory, to diminish the history of a people and a place.

Vancouver was pleased that "the Indian people conducted themselves in the greatest decorum and civility." It's unfortunate he couldn't see that this first contact was a meeting of equals. But the colonial mindset was unable to appreciate the elegance and expertise developed by the Squamish over thousands of years of living on the land, and their deep respect and caring for the waters, their "worship of the Earth and the Rain that wets it."

The Squamish Indian Land Use and Occupancy Study, prepared by Randy Bouchard and Dorothy Kennedy in 1986, quotes several Squamish Nation reports about that first encounter. "The Squamish are said to have considered the Whitemen to be visitors from the dead because of their white skin; their clothes were believed to be their burial blankets." When the Indians were invited aboard the ship, "a Whiteman extended his hand in friendship, but of course the Squamish, unaware of this European custom, believed that the man wished to enter into a game of finger-pulling." It appears the Englishman suffered a dislocated finger, which convinced the Squamish that the ghostly figures were alive and of this Earth, after all.

Captain Vancouver was not overly impressed with Howe Sound, calling it "a dreary and comfortless region." Unlike the Spaniards, who visited the BC coast in the summertime and came up with endearing names for their landing spots—Bella Coola and Esperanza—Vancouver's voyage occurred in the winter. I wonder how he might react to the news that USA Today in their May 13, 2015, issue chose Howe Sound as the number one fantastic fjord in North America. "Between opposing cliffs and more than a dozen islands, visitors come to hike, sail, scuba dive, paddleboard, bike, camp, fish, or experience the triangular fjord from one of many flightseeing tours.... Tourists come from all over the world...primarily for the spectacular scenery and abundant recreation that Howe Sound affords."

As newcomers to what is now called BC began to take over Indigenous lands without offering compensation, they prohibited the Aboriginals to fish and hunt or gather berries and roots in traditional sites. In 1864, Joseph Trutch, commissioner of Lands and Works, the archetypal colonialist, saw British Columbia as a frontier to be developed. He did not believe in Aboriginal title and saw no reason to allow Indians to retain land in an "unproductive condition."

The newcomers attempted to extinguish the language and culture of the Aboriginal people, while removing them from their land. Taking them from their land served to take them from what they considered sacred. This was the first great betrayal of the genius of Howe Sound and its people. Before the Enlightenment period of the seventeenth and eighteenth centuries, when Europe proposed the notion of the universe as a machine and therefore knowable, most Europeans saw the Earth as sacred and mysterious and alive, in common with Indigenous worldviews.

Many Europeans never let go of these beliefs, and a revival of these ideas came together in the Gaia Hypothesis put forward by scientists James Lovelock and Lynn Margulis in 1965. Naming a scientific theory after a goddess did not sit well with many scientists, but the name remained irresistible to Margulis, who theorized that microorganisms had given birth to the Gaian system and continue to be its foundation. More recently, *The Universe Story*, a book written by cultural historian Thomas Berry and physicist Brian Swimme, offers a new narrative. This is a scientific cosmology that also reflects the traditional wisdom of Indigenous people who acknowledge the all-nourishing Earth that has provided for them over thousands of years, and for which they continue to express gratitude. This narrative continues to inform my life.

Amor de Cosmos, who became the second premier of BC in 1872, was born Bill Smith in Nova Scotia to United English

Loyalist parents. After a few years in California, he moved to Victoria, and in 1858 founded the *Daily British Colonist* newspaper, which has survived today as the *Victoria Times-Colonist*. The newspaper covered the smallpox epidemic of 1864 in Victoria. It called on the government to eject those Aboriginals most badly affected by smallpox out of the city. Without any help or aid, the infected people were expulsed by the government to all parts of BC, causing a province-wide epidemic.

De Cosmos actively contributed to the anti-Chinese and anti-Aboriginal sentiment prevalent in popular and political discourse in the province. Hotheaded, flamboyant, and before his death in 1897, declared of unsound mind, he indulged in vicious hate mongering towards Aboriginal peoples in the province. "Shall we allow a few vagrants [the First Nations] to prevent forever industrious settlers from settling on unoccupied lands?"

For the colonialists, the land and what was on, under and around it, was there to be taken and used. This allowed them to overlook complex systems that were already in place, systems that were helping to maintain and support resources rather than use them up or waste them. This belief that the land "needed to be improved" was in sharp contrast with the economy the Squamish had developed, one that was finely tuned to the land and the waters. Unfortunately, this colonial attitude persists as corporations and politicians invest in grand gestures to make the land "productive."

Tame the wilderness. Break the land. No way to treat your Mother. Nature is not something to be conquered. Not something to be brought to its knees. Nature is something to be grateful for. An Elder to respect, to learn from. Nature is the first and best teacher.

In 1870, the colony of British Columbia unilaterally denied the existence of Aboriginal title, claiming that Aboriginal peoples were too primitive to understand the concept of land ownership.

"This rationalization," writes Alex Rose in his book *First Dollars*, "along with the myth that BC was an empty land, was all that was needed to undertake the largest alienation of property, half the size of Europe, in imperial history outside of Australia." He calls this land grab "theft, plain and simple."

This land grab was codified in the Indian Act passed by the federal government in 1876, making Aboriginal people wards of the state, dependent on the Indian agent, and unable to leave their reserve without permission. The Act defined who was an "Indian," and was especially hard on women. Native women who married non-Native men were no longer considered "Indian," nor their children, although non-Native women who married Native men were. If an Indian woman married an Indian man, her status was with his reserve. If he died or left her, she lost her status and could no longer live on the reserve. Indians were unable to get a loan or a mortgage, start a business, fish or hunt in their traditional territories, sell their land or market their skills. Unless they left the reserve and severed their ties to their community, they were not allowed to vote. They could also lose their status if a grandparent fought in the war.

The Indian Act exists today—with some modifications—and continues to legislate many First Nations into poverty. Although bands were given reserves to live on—often on poor and small amounts of land—they don't technically own the land. The legal title is vested in the Crown and held in trust. This trust was broken again and again as Native land was "needed" for development.

The unique systems of governance established by the First Nations were also affected. The imposition of an electoral system further undermined a system of self-governance that had functioned effectively for thousands of years. Although band councils, as decreed by the Indian Act, were elected by band members, they were accountable to the minister of Indian Affairs. There was little understanding of Native social organization on the west coast, or that each nation had its own style of governance. By 1880 each Squamish reserve had a chief, whether elected, chosen by consensus or appointed. The colonial presence continued to redefine maps, people, cultures, religion, imagery and ways of seeing and believing. "The Indian Act destroyed decision-making," says Russell Diabo, a Mohawk educator. "Decisions were made for us." This contributed to a breakdown of the social fabric of Indigenous communities.

The natural beauty of Howe Sound and lush farmland attracted newcomers. The first settlers arrived in 1888. They appreciated the mild climate and the abundance of food. They worked hard and coexisted with the Squamish Nation and sometimes intermarried. But with the arrival of the newcomers, the Squamish experienced a tsunami of changes to their way of life. Disease had already begun to ravage the Indigenous population in BC—most dramatically smallpox, but also measles, scarlet fever, influenza, venereal diseases and tuberculosis—leaving behind empty villages. Bob Baker, an Elder who lives in the Squamish Valley, told me he once asked his great-grandmother, Ma Baker, why she married an Englishman instead of a man from the Squamish Nation. Her answer was heartbreakingly simple. "All the men were dead." From an estimated population on Canada's west coast of one hundred thousand at time of contact in the 1770s, there were twenty thousand left in the 1920s.

The story of Howe Sound changed dramatically with the arrival of the Europeans and their new technology, guns, whiskey and new diseases. The Europeans also imposed a new religion and a new way of using the land. To facilitate the homesteading of land by settlers, the Squamish were placed on twenty-four reserves. They were promised better lives, and jobs in the new economy, but the very projects they worked on would ultimately impair their ability to maintain their own cultures. The building of dams, dikes and river diversions eliminated or restricted salmon runs; and the cutting and then hauling of giant logs through fish spawning channels further harmed the habitat of the fish and the plants they depended upon for their existence.

The conversion to Christianity helped to erode the Indigenous sense of identity and language. Their ability to pass on

Indigenous knowledge was curtailed by language restrictions and the loss of control over their land. Since women were the main transmitters of knowledge about plants, medicines and food processing, the Indian Act had a devastating impact on opportunities for passing on this vital information.

Most disruptive to the transmission of knowledge were residential schools, often in faraway places, where young people were separated from the wisdom of their Elders and forbidden to speak their traditional language. An estimated 150,000 Indian, Inuit and Métis children between the ages of four and sixteen attended residential schools, which operated in seven provinces from the 1870s to the 1990s.

Prime Minister John A. Macdonald promoted the need for residential schools in a contemptuous speech to the House of Commons in 1883: "When the school is on the reserve, the child lives with his parents who are savages, he is surrounded by savages and though he may learn to read and write, his habits and training and mode of thought are Indian. He is simply a savage who can read and write."

One has to take but a cursory glance at the Truth and Reconciliation Commission Report to appreciate the full horror of these schools and how they gave the state, through the churches, total control over future generations.

In conversation over a bowl of soup, Squamish Nation Elder Gloria Nahanee shares the devastating effects of "Indian schools" with me. Her father had attended the residential school in Sechelt, but he rarely spoke about his experience. "He was not affectionate to me." Only near the end of his life did she begin to understand his pain. "My father and I were in a coffee shop," she says. "It was around Christmas time. He had poured cream and sugar into his coffee and he kept stirring the cup so violently that the coffee was spilling all over the table. I had never seen him act that way and I asked him what was wrong." I watch Gloria's beautiful, expressive hands as she demonstrates her father's frantic stirring. She leans across the table and explains: "There were Christmas carols playing in that restaurant," she says and pauses.

Then Gloria reveals what her father had carried all his life and never shared with her until that day. "They played carols at the school," he told her, "but they wouldn't let us go home for Christmas." We are both silent for a moment as we take in the heartbreak of many of the children who attended residential schools.

"We were degraded and humiliated," Gloria adds, speaking of her own experience in the schools. "When I was young, I didn't learn about who I was or where I came from. They wanted to break our spirits." At age nineteen, she began to study with her Elders, to travel with them, to learn the ways of her people, the way her ancestors did everything in a sacred manner. She also learned the difficult art of self-respect and self-love. Love was missing, she tells me. "My ancestors were wonderful—on the mainland, on the Island and in eastern Washington. They would bring out the bannock and tea. Learning from my Elders changed my life."

For years Gloria has played an important role in her community, doing ceremony on weekends, and during the week working with youth at risk so they might also find respect for themselves and their culture. "All I'm doing is sharing the good things, the teachings. I share this with the young ones and make a difference." Gloria was also instrumental in re-establishing pow-wow dancing and has seen how the young who have participated have flourished. Her daughter, Kinani, has danced since she was young. "She is not afraid to use her voice." Forbidden to speak the Squamish language, Gloria is now taking classes to learn the language of her ancestors. "It's not easy. So different from English. My ancestors must have been really smart to be able to speak that difficult language."

Gloria also shares with young people what happened to her in the schools. "The young people are thanking me, telling me now they know what their parents and grandparents are going through. Our people are still suffering." She describes a powerful exercise that she recently participated in with other Elders and young people at a local school. The ages were grouped in concentric circles—an inner circle of young adults, then outer circles of middle-agers, Elders, nuns and priests. In the centre were the children. Then the children were removed. What was left was a village without children. No children to hold, to cuddle, to enjoy, to educate, to love. No children to parent. No children to learn the language and the culture of their people, no children to learn how to be parents themselves. This is what the village looked like for up to seven generations. Apologies, yes, but much more is needed to make proper restitution for residential schools, and the historical trauma inflicted on First Nations people throughout Canada.

In 1884, because of increasing pressure from the missionaries, the Indian Act was amended to outlaw cultural and religious ceremonies, such as the potlatch—the major social, economic, political and spiritual institution of First Nations on the west coast. The word "potlatch" is a Chinook jargon word that means "to give." The ceremony was an important cultural event for the distribution of wealth and the witnessing of ceremonies, such as a marriage, or a young person's entry into adulthood. It was also the ceremony to announce the appointment of public positions. Guests were feasted and given gifts—many of high value—in appreciation for their attendance as witnesses. The celebration was based on an economy that encourages the movement of goods and wealth rather than hoarding for one's own gain.

The potlatch was a primary way to communicate, to bear witness, and to keep peace between families. To the European mind, the potlatch appeared to be a lavish waste of wealth. John A. MacDonald called it "debauchery of the worst kind, and the departmental officers and all clergymen unite in affirming that it is absolutely necessary to put this practice down." New laws were imposed to ban the spiritual practices of Indigenous people; putting on a potlatch was subject to imprisonment and the confiscation of property. Although a few people continued to hold them in secret, many of the cultural songs and dances encoded with traditional knowledge were lost.

John Ralston Saul reframes the banning of the potlatch in *The Comeback: How Aboriginals are Reclaiming Power and Influence*:

"On the West Coast, the banning of the potlatch ceremonies was dressed up as a reform to protect the economic wealth and moral well-being of the natives. It was actually an attempt to weaken those natives who had adjusted successfully to the newcomers' economic system and made themselves powerful players in the new fishing industry." The Squamish also lost the right to make a living from what they did best: fishing. In 1889, the federal government excluded all Aboriginals from commercial fishing, forcing them to work in canneries and logging operations, taking down the trees that for many of them were Elders, ancestors and teachers.

They were also being removed from the reserves that had been promised to them in order to facilitate European settlement. Before the opening of Stanley Park, there were several Squamish settlements across the Stanley Park peninsula, the largest of these known as Xway xway—pronounced *whoi whoi*. It's estimated that they inhabited the area as far back as three thousand years ago. The longhouse, in what is now called Lumberman's Arch, was sixty metres long and twenty metres wide with eleven families and one hundred adults and children. Most of the inhabitants were unceremoniously removed, their gardens, fences and properties vandalized and destroyed. They paddled across the inlet to North Vancouver and relocated there. They received no remuneration.

In June 13, 1886, as the Great Vancouver Fire burned, the Squamish women who had been relocated to North Vancouver saw people floundering in the waters of Burrard Inlet. They risked their lives to paddle across the inlet, pick up survivors and bring them to safety. The fire destroyed most of Vancouver's buildings and dozens of people died. To keep their spirits high in this dangerous undertaking, the Squamish women created a canoe song, which continues to be sung today. The Woman's Paddle Song has resonances of hymns the women sang at Sacred Heart Church, and each time I listen to it, I remember this important piece of long-neglected history and my heart melts.

One can only imagine how the women felt, having been recently evicted from their homes, and yet determined to save the lives of the settlers on their land. The depths of compassion shown by these women is the most compelling example of forgiveness and reconciliation I have ever heard. May we all be as open and willing to save the lives of others, even of those who have betrayed us.

The history of the Squamish Nation continued to be extinguished and removed from Vancouver history. In her paper "Erasing Indigenous Indigeneity in Vancouver," presented to the Canadian Historical Association in May 2007, Professor Emeritus Jean Barman writes about the "final erasure of indigenous Indigeneity from Vancouver." In 1923, the Vancouver Parks Board put up four Kwakwaka'wakw totem poles, "the replacement of indigenous Indigeneity with a sanitized Indigeneity from elsewhere."

The Squamish village of Senákw was located on the shore of what today is called Kitsilano. Just over one hundred years ago, the city of Vancouver was a lush landscape of forests, swamps and salmon-bearing streams. In 1913, the BC government, without federal authority, pushed through an arrangement to sell the seventy-two acres known as Kitsilano Reserve No. 6 to the Canadian Pacific Railway for a real estate development. The Squamish were evicted from their reserve and put on scows, some going to the North Shore, others up Howe Sound. When they looked back, they saw their houses burning in the distance. This eviction happened during the winter, and those who travelled to Squamish without their usual supplies of stored food barely survived.

The Squamish were determined to get back their rights to fish and hunt. That was their livelihood. In 1906, Chief Joe Capilano, whose ancestral name was Sa7pelek, led a delegation of BC chiefs to London to speak to King Edward VII. Chief Joe was born in the 1850s to a high-status family in the Squamish Valley. He was a respected carver and spirit dancer. A strong, healthy man, he climbed to the top of Chíchiyuy—the Sisters—to learn about peace making.

Before he went on this historic voyage, he travelled around BC for two years to talk to all the First Nations and get their approval. That way, when he met the king he could say he carried the handshakes of all Indians in BC. The chiefs—all twenty-three of them—left on July 3, 1906. It was a gala event with speeches, a brass band, hundreds of Indians and hardly a dry eye as the Imperial Limited pulled out of the Vancouver train station. The chiefs were all the rage on the streets of London, wearing their finest regalia. Joe Capilano loved London and London loved Joe. They took photos of him as he toured the city. When he saw Hyde Park, he said it would make a good hunting ground. Something he was denied in his own country.

The delegation of chiefs presented their petition to the king. They wanted to be able to fish and hunt in their traditional territories. They wanted the ban on the potlatch removed, and they wanted compensation for lands that were stolen from them. They weren't asking for new protections. They wanted the royal promises that were made under Queen Victoria in the 1763 Royal Proclamation to be honoured: "...that the several Nations or Tribes, with whom we are connected, and who live under Our Protection, should not be molested or disturbed in the Possession of such parts of Our Dominion and Territories as, not having been ceded to, or purchased by Us, are reserved to them, or any of them, as their Hunting Grounds...and We do hereby strictly forbid, on Pain of Our Displeasure, all Our Loving Subjects from making any Purchase or Settlements whatever, or taking Possession of any of the Lands above reserved, without Our Special Leave and Licence for that Purpose first obtained. And we do further strictly enjoin and require all Persons whatever, who have either willfully or, inadvertently seated themselves upon any Lands within the Countries above described, or upon any other Lands, which, not having been ceded to or purchased by Us, are still reserved to the said Indians as aforesaid, forthwith to remove themselves from such Settlements."

The English king listened carefully to the chiefs. He agreed that Aboriginal property rights in the land should be restored but only through the Canadian government. There was just one problem with that idea. Indigenous peoples weren't allowed to vote. They would have to wait more than fifty years for that. With no vote, they had no way to influence the government. Capilano was a great leader, a visionary. Today he has a university named after him and a ferry that takes passengers from Horseshoe Bay to Bowen Island/Kwilákm. But in 1906 he returned home with a broken heart.

In a 2005 paper, "Rethinking Dialogue and History: The King's Promise and the 1906 Aboriginal Delegation to London," Keith Thor Carlson expresses his admiration for Joe Capilano's London trip. "By bypassing the British Columbian and Canadian governments, Capilano and his associates would come to represent for non-Native society, a new generation of westernized, practical leadership. Yet, such rational behaviour did nothing to diminish the fact that in Indigenous eyes, Capilano remained a symbol of continuity with the spiritually potent world of their ancestors." The important teaching for non-Aboriginals was that Indigenous people are not going anywhere. They have no homeland to go to. This is their homeland, and they're staying. The delegation was an important step for Native people, the beginning of a process of taking political protest into their own hands, one that continues, even more powerfully, to this day.

The practice of Indigenous stewardship continued to be replaced by the European ideology of extraction and profit, whether it came to furs, fish, trees, gold or humpback whales—which were hunted to extinction for their oil by the early 1900s. The Gold Rush, which overtook British Columbia in the late nineteenth century, was a bit of a crawl in Howe Sound. There was plenty of shiny yellow stuff in the Squamish River and several claims were registered, but it was mostly iron pyrite—fool's gold. Copper ore was discovered in 1888 in the mountains around Britannia Creek, south of Squamish on the eastern shore of Howe Sound, and large-scale mining began at Britannia Beach in 1905.

By 1929, Britannia Mine became the largest producer of copper in the British Empire. More than sixty thousand people lived in the mining community over the seventy-year life of the mine, during which fifty million tonnes of ore were extracted.

While there were economic benefits for both the local community and the province, the environmental cost was high. Every day, hundreds of kilograms of dissolved heavy metals such as copper, iron, cadmium and zinc leached out of the mine and flowed directly into Howe Sound. Countless millions of gallons of effluent laden with toxic chemicals were discharged into the Sound for well over half a century.

Although the mine closed in 1974, it left behind a poisonous legacy. Acid mine drainage, the toxic result of mining, is like atomic waste. Unless actively treated, it will continue for thousands of years. Snowmelt and rainwater continued to flow through the mine's abandoned tunnels, bringing the toxic metals with it and contaminating the waters. For close to a century, Howe Sound had to deal with as much as 450 kilograms of toxins daily. The clarity of the water was an indication that no living creature, no fish, no insect could survive in Britannia Creek. Britannia, once a river full of life, was dead.

In the early 1970s, government biologists found that the impacts stretched north to the Squamish Estuary, affecting millions of juvenile chinook salmon. Polluted runoff from the site was believed to jeopardize more than four million young salmon every year. In 2001, Britannia Creek was designated by the Outdoor Recreation Council as BC's most endangered river, and the area extending into Howe Sound was considered the "worst point source of mineral contamination in North America" by Environment Canada.

The mine had a succession of owners, some American, some Canadian, and the issue of who should pay the cleanup costs dragged on for years. To add insult to injury, the owner, Copper Beach Estates, had plans to turn the former Britannia mine, an open pit nine hundred metres above Howe Sound, into a contaminated dumpsite—an extreme example of "if you can't fix it, feature it." Fortunately, this outlandish proposal was not realized.

In 1909, the Port Mellon Mill on the west side of Howe Sound near Gibsons produced the first wood-fibre-based paper in

BC. Although the mill provided jobs for locals, it was a major source of air and water pollution. People as far away as West Vancouver reported the nasty sulfurous smell. Then in 1912, the Woodfibre Pulp Mill opened near Squamish where Mill Creek empties into Howe Sound, on a site known as Swiy'a'at by the Squamish Nation. More toxic chemicals and metals were allowed to enter the waters. Organochlorines, such as dioxins and furans—generated when elemental chlorine is used during the pulp mill bleaching process— were discharged as effluents. Although new efficient systems were later installed, the effect on Howe Sound was disastrous.

In 1965, a chlor-alkali plant was built at the head of Howe Sound by FMC Canada. The chemical company employed approximately seventy-five people and produced 175 tonnes of chlorine, 200 tonnes of caustic soda—sodium hydroxide—and 30 tonnes of muriatic acid—hydrochloric acid—daily, becoming the second largest polluter in Howe Sound. Squamish Nation Elders reported that there were no more berries on the bushes. The air was acrid. From 1965 to 1970, as much as 40 tonnes of mercury may have been released into Howe Sound. The toxic mercury levels meant the Squamish people could no longer eat the fish they had thrived on for thousands of years. Mercury never really breaks down; it becomes more concentrated as it moves up the food chain.

Pollution from pulp mills, the chemical plant, mining, clearing for development and agriculture, and the disruption of hydrology from logging and road building, combined to degrade the quality of life in the waters of Howe Sound. The flourishing commercial prawn fishery closed, and then the crab fishery. Formerly productive salmon-bearing streams became too polluted or blocked for returning salmon.

But the biggest threat to salmon has always been politics. Over the years, salmon, the iconic and foundational fish of the west coast, has been considered by powerful political interests to be less important than timber, pulp mills, railroads and mining. Today, the threats are multiplied: a gravel mine, pipelines, supertankers and LNG. The disappearance of salmon on the west coast is on the scale of the extermination of the bison on the plains. It signals not only the loss of an economy but the end of a way of life.

People who live in the Squamish area have needed to be resilient, as well as in possession of a good sense of humour. A poem by Andy Olson published in the *Squamish Times* on March 1, 1972, is testament to that.

> Squamish we love you, you're our kind of town
> Nestled away at the end of Howe Sound.
> Your mountains are white, but your rivers run mud
> Your end of the sound is all covered in crud.
> Yes there's flotsam and jetsam from the log booms set free
> But without these bush barons just where would we be?
>
> Your future looked great with Brennan and Shrum
> We gave them the tools, now look what they've done.
> Your Harbour's expanding, still waters run deep
> We've chlorine and mercury to lull us to sleep.
>
> Your delta was beautiful and the fishing was swell
> Now thanks to your gravel, we'll bid them farewell.
> Your chemical plant has a wondrous phew
> But if one stink is good we must surely have two.
>
> Woodfibre is near, Britannia as well,
> They are part of a plot to make our life hell.
> But give us some sun, a wonderful day,
> Our town is so beautiful most of us stay.
> Yes, Squamish we love you, you're our kind of town
> Nestled away at the end of Howe Sound.

Elsewhere in Howe Sound, there have been many proposals for industrialization, often meeting with resistance. In 1979, Gambier Island residents gathered to protest a mineral company's plan to mine 252 million tonnes of copper-molybdenum, a project that would include the construction of several dams, a large tailings pond and huge rock storage piles. The entire project would have encompassed two-thirds of the island. Fortunately, the Islands Trust, a federation of local island governments with a mandate to make land use decisions that will "protect and preserve," stepped in. Two trustees, Elspeth Armstrong and Beverly Baxter, led the crusade to oppose the giant open-pit mine. After five years of petitioning government officials, the project was shelved and the provincial government established a mineral reserve on the entire island in 1985.

In the early 1900s, Tunstall Bay on Bowen Island/Kwilákm was the site of a dynamite factory. Western Explosives hired eighty employees, including Chinese and Japanese workers. After several explosions and the deaths of eleven men, the plant was shut down. Other Howe Sound disasters include the many boats that succumbed to the Sound's depths. Howe Sound's deep waters have been the graveyard of many ocean-going vessels, many of them tugboats that burned, sunk or were beached throughout the Sound.

In *The Rush for Spoils: The Company Province 1871–1933*, published in 1972, Martin Robin, now professor emeritus at Simon Fraser University, takes a critical look at the narrative of resource extraction and "the pioneering acquisitive culture" that has existed in BC since colonization. In the late 1800s, writes Professor Robin, "the Coast province stood nervously poised on the threshold of the development of a massive resource extractive economy, destined to be carved out of the majestic wilderness by rapacious entrepreneurs acting through companies." All of it made possible by a few treaties and mass expropriation of Indigenous lands, which he refers to as "legalized robbery."

The spirit of the province, he writes, was "speculative, acquisitive, adventurous," and it attracted a colony of dreamers drawn to BC like a magnet: "the California miner who hoped for ease and wealth…the Oriental who longed to escape blighting rural poverty…the retired colonial official or delinquent son of the British gentry who wished for the calm of a pretty green estate in the colonies…the buccaneer capitalist who lusted after the fortunes hidden in the mountain wilderness. This was no community which formed, but a cauldron, a mass of men, jostling, shoving, and colliding in the pursuit of profit."

Robin describes how politicians promoted various industries as the salvation of the province's economy. "At one time, Cariboo was to 'make the country' … then lumber was to be the 'fortune' of everybody, then copper, then coal. Gold and sometimes silver have been the materials on which the often-sanguine colonists hoped to hold on." Add whale oil and cedar to the list and you get the picture of what was going on in Howe Sound. Each of these extractions in turn propelled the cycle of fevered boom, spinning downward to a lukewarm bust. And the devastating damage and costly cleanups that followed.

Robin is blunt when it comes to exposing the myths of the frontier mentality. "The new mythology propagated by the politicians, publicists, and company lords was raw development, an ideology romanticized by the concept of the golden frontier. A lady-in-waiting with great expectations, British Columbia blushed and palpitated with each new thrust into its interior. [!]" (My exclamation mark.)

Professor Robin's book was published over forty years ago. I wanted to know what he thought about the current state of industry and development in BC and if anything had changed. So I tracked him down in Vancouver, and he came out of retirement briefly to speak with me. "There are some constants," he said. "The inequality, the adulation of growth, big money coming from outside. The worship of growth is an enduring theme." One aspect that has changed is the "powerful movement worldwide and in BC of people working for the protection of the environment. Climate has moved to the top of the agenda and that's new."

When I asked how we might move from the ideology of growth to something more sane and sustainable, he said it has to come from government. "The role of the state should be to protect and preserve the life of its people, to be sensitive to, and influenced by, more groups than just big business." In those frontier days, there were no adequate environmental measures associated with these important industries; government abandoned its role as protector of the common good. Companies were given free reign. They were not bound to strict environmental laws, not legislated to maintain the safety of their operations, nor to clean up when things shut down.

The fox was managing the henhouse. The colonial mindset could not accept that nature is not "out there" but in us, that air, water, soil, forests and oceans are our shared wealth, our shared responsibility. And we are only as healthy and wealthy as the land, as the many species that provide beauty and sustenance. I find it disconcerting that while BC is blessed with remarkable wildlife not found anywhere else in Canada—killer whales, Grizzly bears, barn owls and badgers—there is no provincial endangered species act.

A *Vancouver Sun* editorial, "BC Economy about More Than Just LNG," in the June 10, 2016, issue, sheds some light on the major industries in BC. "The province's economy," it states, "has evolved far beyond its muscular, now romanticized roots in the resource sector when loggers, miners and fishboats symbolized prosperity." Although the editorial is supportive of LNG, statistics on the sectors providing the most jobs are a bit of a surprise, given the provincial government's focus on job creation in oil and gas.

"For every oil and gas worker, there are almost three people employed by the motion picture industry…three in the performing arts, three in the publishing trade. Four workers in agriculture, four in forestry for every one in the oil patch. Seventeen high tech workers and twenty-five tourism workers. Professional, scientific, and technical workers in BC now outnumber oil and gas workers by a ratio of thirty-six to one. In construction forty to one, and in retail fifty to one." Closer to home, the Woodfibre LNG plant, if it gets built, will provide one hundred full-time jobs, while the Sea to Sky Gondola employs 250 people. Seems the time has come to let go of the old frontier myths and embrace a BC economy that offers new opportunities for making a living without spending the natural capital of the province.

I looked for current stats on the number of tourists who visit the Sea to Sky area. To give you an idea, the gondola, which opened in the spring of 2014 just south of Squamish, welcomed its one-millionth visitor in January 2017. Not all of these visitors were tourists; locals also enjoy the spectacular views and the hiking. In 2015, the Squamish Visitor Centre saw more than fifty-nine thousand tourists walk through their doors, up 30 percent from the previous year. And the Vancouver Aquarium's Coastal Ocean Research Institute (CORI) Report on Howe Sound estimates there were over two million visitors to Howe Sound Parks in 2015, a 48 percent increase compared to 2010.

Not to be undone, Whistler, fifty-eight kilometres north of Squamish on the Sea to Sky Highway, is home to Whistler Blackcomb, one of the largest ski resorts in North America, and host to the 2010 Winter Olympics. A town of less than ten thousand inhabitants, Whistler welcomes over two million people a year to ski and snowboard and snowshoe the mountains that feed and drain into Howe Sound. Whistler has become a year-round destination featuring sports, arts and cultural activities, and events. The Whistler Writers Festival attracted over twelve hundred participants at the fifteenth annual festival in 2016.

The Audain Art Museum, nestled quietly in the forest, opened in 2016. A 43.5 million dollar gift from philanthropist Michael Audain, it features the art of BC from the traditional works of First Nations through to its contemporary masters. The permanent collection includes the world's most important collection of northwest coast masks and two dozen Emily Carr works. In its first year, the museum attracted fifty thousand visitors.

The Squamish Lil'wat Centre in Whistler, a popular destination, is a showcase for the art, history and culture of the Squamish and Lil'wat Nations. Visitors are greeted with a welcome song, taken on a guided tour and offered the opportunity to make a traditional craft. The artwork alone is worth the visit. The centre is a model of collaboration, two nations coming together to create something of great beauty and lasting value.

The influence of First Nations is one of the most remarkable aspects of change today. They are taking a much more prominent role in the affairs of business and environmentalism. Growth is not a part of their traditional ideology or worldview: they favour the protection of what supports life. In his book *The Wayfinders*, anthropologist Wade Davis reflects on "this particular attitude of ours, this manner in which we have reduced our planet to a commodity, a raw resource to be consumed at our whim…there are in fact many other options, any number of different ways of orienting ourselves in space and landscape." Part of my journey of discovery was to search out these different ways of orienting ourselves in landscape.

The Squamish Nation is the only government in Howe Sound looking at the Sound as a whole, intertwined ecosystem. I was fortunate to be introduced to Randall Lewis, environmental advisor for the Squamish Nation and president of the Squamish River Watershed Society. Randall grew up on a Squamish reserve in the 1960s: "chasing cougars and bears up and down the Cheekye River," he tells me over lunch at the Howe Sound Inn in Squamish. His father, Pel'Wilem, left residential school when he was thirteen, moved to Seattle, and later became a fisherman, hunter and logger.

Randall tells me some of his people went to Seattle because in Canada they were not allowed off the reserve without permission. His mother, Maakwa-Aalth, a Nuu-chah-nulth from Port Alberni, ran away from residential school several times. The Elders hid her in the attic but she was found and returned to the school. She ran away again and went to the US, where she met Randall's father at a Glen Miller Dance in Seattle. Randall is part of the Wolf Clan, descended from a man who was raised by wolves near the Cheakamus River. He proudly carries the ancestral name he was given by his father, Ta'haxwtn. It means "high spirit of the headwaters of the land."

For the Skwxwú7mesh people, receiving an ancestral name is a significant event. The Elders watch the boy or girl carefully, looking for strengths and similarities to an ancestor. The name will be a model for the child, an inspiration, something to live up to. When the name is chosen, there is ceremony with drumming and dancing and feasting. And when the child receives that name, it means they will take on, in some way, the personality and skills of their namesake. Their ancestral name is like a beautiful blanket, something to wear proudly, to take care of. The child is told to look after that name. Don't do anything to tarnish it. Carry it well.

As a boy, Randall was taught by his Elders, some over one hundred years old. They told him stories, the same stories, over and over again. When he asked them why they repeated the same stories, they told him they wanted to make sure he would never forget the history of his people. Don't forget what we're telling you, they said. You're going to carry this. "This was a blessing to me as a young boy," he tells me.

In high school, he "worked very hard for his Bs and B pluses," and when he was asked to choose a language to study, he chose Squamish. The school insisted that this language didn't exist. How could that be possible, he countered, when my Elders speak it every day. He repeated once again that he wanted to study Squamish. "They kicked me out of school because I kept saying I wanted to study Squamish." In desperation, the school administration called a conference and Randall chaired a meeting with the school principal and Elders. This meeting triggered a research process to start gathering information for curriculum development, and today the Squamish language is offered at Don Ross Secondary and is taught by a Squamish Nation language teacher. Students applying to university can use Squamish as their second language.

Randall remembers from his childhood the enormous runs of eulachon, a small fish from the smelt family, almost as important to coastal people as salmon. It was the first fish to arrive in rivers at the end of long wet winters, a salvation for hungry people. It was rich in fat and could be dried and lit like a candle, which is why it was also known as candlefish. The grease—used as a condiment, salve, seasoning, preservative, laxative—was a rich source of vitamins A, C and E. "There were no laws back then," Randall says, "and there were creosote logs right on the shore into the Mamquam Blind Channel. Fifteen-feet deep of mercury, arsenic, furans and dioxins flowing from the chemical plant into the gutters, and fifteen-feet deep of herring and salmon and eulachon carcasses."

He also remembers a football practice in 1973, and the sound and smell of the explosion at the chemical plant. "Sirens going off, the wind blowing hard, orange-green clouds flowing toward us—toxins, chlorine clouds." He ran into the school but the clouds followed. There was no safe place. After the explosion, the Elders advised them not to eat the berries and the plants on the land.

As a young man, he admits he was "all blood and guts, ready to go, arrows a-blazing, sue the governments for ignoring their fiduciary obligations." It was the late Chief Joe Mathias who told him: "I hear what you're saying. You need to back up. It took us a long time to build these foundations and principles. We can't burn those bridges." Randall understood that it had taken thirty to forty years to create a dialogue with governments. It was more important to build collective community consensus.

Randall has witnessed the poisoning of Howe Sound and the cover-ups. On August 5, 2005, a CN train en route from North Vancouver to Prince George derailed in the Cheakamus River canyon near Squamish. Nine cars derailed, including one carrying caustic soda. The effect on the fish was devastating. "Young trout and salmon were just re-entering the river," says Randall. "I'm watching fish jumping out of the river, their eyes burning from the caustic soda." Half a million salmon were killed.

"When I was studying forestry and business resource management at BCIT, I came across a 1978 report that had been buried by DFO, Department of Fisheries and Oceans. It was a study of the effects of the pulp mills at Port Mellon and Woodfibre. The study showed there were seventeen parts per million of dioxins and furans in the tissue of young chinook salmon in Howe Sound. Later studies showed the presence of dioxins and furans in the pancreas of crabs. Ten years later, DFO shut down all of Howe Sound to crab and shrimp fishing. But some Elders continued fishing and eating because there was no other choice and that's all they knew."

Randall is a burly man, sturdily built, one of fourteen brothers and sisters. He has the focus and steadiness of a warrior. He tells me with pride that the Wolf Clan is a clan of warriors. He has worked as a fisherman and as a faller doing helicopter logging. He also worked for the federal government in the Immigration Department and sits on the board of various organizations, although the word "sits" doesn't describe the energy and dedication he brings to every meeting and every project.

He's a great talker, and fast, the words tumbling out as I try to take notes. He speaks passionately about "the race for the spoils in our territory" that took place during the nineteenth and twentieth centuries, and continues to this day. "This is not our way. Our sacred oath is whatever you take and use you return threefold," he told me. "We're dealing with a history of colonization. Take, take, take: furs, gold, forests, fish. We have four thousand members in sixteen tribes. We can't live off our resources anymore."

The provincial and federal governments, "by cutting appropriate legislative due diligence to protect habitat, have given industry direct access to approvals with weakened regulations," he says. "The result is unintended consequences for our children and our environment." At the same time, "the federal government, by bundling omnibus bills together, has neutered DFO. They effectively have no mandate anymore. Governments of the day have no meaningful accountable mandates for fiduciary obligations.

The Navigable Waters Act and Species at Risk Act were also significantly marginalized to accommodate industry." As a result, the Squamish Nation has been forced to sue governments to live up to their obligations to protect the land.

For Randall, "There's only one thing left in our war chest to fight the Crown: Section 35 of the Constitution of Canada." Section 35 recognizes and affirms Aboriginal rights—rights that existed long before the Constitution of Canada was drawn up. "We have to use this section all the time to defend our inherent Aboriginal rights and title," he says. These are the same rights and titles that Chief Joe Capilano defended over one hundred years ago. The same rights and titles that Andrew Paull, a self-taught constitutional lawyer, fought for in the twentieth century.

A visionary leader, Paull spearheaded the amalgamation of sixteen Squamish-speaking tribes, united together under the governance of the Squamish Nation. The amalgamation was signed on July 23, 1923. The purpose was to draw strength by coming together and to reclaim the land that had been stolen from them. The amalgamation seems to have been successful. In June 2000, the Squamish Band reached a 92.5 million dollar out-of-court settlement with the federal government settling claims to former reserves in Kitsilano, North Vancouver and Squamish, and more than a dozen other parcels of land. And a BC Court of Appeal decision ordered CPR to return eight acres of the Kitsilano land, the only bit of land that wasn't built on. Colonization didn't end in the nineteenth century. The ideology is still at work today. Land is no longer a living gift to be appreciated, respected and protected; it has become real estate, a commodity to be bought and sold.

For thousands of years, Howe Sound was a powerful force of nature, increasing in beauty and fertility, caring for her spawning fish, her free-running waters, offering the sumptuous buffet of her estuaries. Until it came to an abrupt end. The cumulative impact from the pulp mills and the abandoned copper mine left the waters of Howe Sound largely lifeless. Her waters had reached the limits of their ability to absorb the various industrial toxins. Howe Sound was in distress. Her ability to nurture life had been diminished. Locals speak with a heavy heart about the time when the eulachon and the whales disappeared, and the herring and the salmon were seriously diminished. By the 1980s, large parts of Howe Sound were considered biological "dead zones."

You writers. All the same. You love going on about the doom and the gloom. Sells newspapers, they tell me. And books. But you should never underestimate the fierce power of the breathing wild, its ability to endure, to persevere, to pull itself up by its roots, to triumph.

THE RENEWAL

On January 1, 2013, Jonathan Bell, sailing enthusiast and commodore of the Bowen Island Yacht Club, entered his boat, the *Takaya*, in the annual Polar Bear Race in Howe Sound. Everything from America's Cup classic sailboats to old wooden ketches would attempt to complete the twenty-five-nautical-mile race in an average of five hours. "The race got off to an interesting start," Jonathan reports, "with the wind shifting in many directions" and he and his crew scrambling to work with every gust. Then suddenly, the winds died down, and wonder of wonders, the dolphins appeared, playing under the bow of his sailboat. "It was unbelievable. Every boat was being escorted by a pod of white-sided dolphins. I've never seen anything like it. We were so excited, we didn't even mind not winning the race."

Bell and his crew were understandably excited. Dolphins had not been seen in the waters of Atl'kitsem/Howe Sound for over fifty years. Recently, humpback whales have been filmed in the waters of the Sound by kayakers, thrilled at the sight of the massive mammals breaching a few feet from their boats.

The story of the renewal and transformation of Howe Sound is nothing short of a miracle, a miracle sponsored by committed volunteers, governments and corporations working together. This recovery from a legacy of industrial misuse didn't happen overnight. Although the Squamish Nation had been fighting for decades to change things, it took time for other residents of Howe Sound to wake up to the fact that their waters had been poisoned, their air polluted, and their home had become a garbage dump. They began to organize and protest. They formed local and regional groups and addressed their concerns to the governments of the day.

A major healing process was needed to overcome years of unsustainable development. The Save Howe Sound movement was formed, putting pressure on governments and corporations to clean up the air in the Sound. Terry Jacks, singer, songwriter and creator of the popular song "Seasons in the Sun," became the poster boy of the campaign and one of its major campaigners. The effort was successful. In 1988, Howe Sound Pulp and Paper began a 1.3 billion dollar renewal process at their Port Mellon mill to turn it from an environmental disaster into one of the cleanest pulp mills in the world. The Woodfibre Pulp Mill closed in 2006. The air in Howe Sound became breathable once more.

It took thirty years to stop the runoff from the Britannia mine, but Mark Angelo, conservationist and founder of International Rivers Day, refused to give up. In the 1980s and 1990s, he led a sustained public awareness campaign, which turned the tide. In December 2001, UBC mining engineers placed a huge concrete plug, the Millennium Plug, into the main mine to redirect the acid mine drainage, a consequence of mining activities. The goal was to contain the pollution of an upstream tributary of Britannia Creek. In 2005, the provincial government and industry joined forces to finance the construction and operation of a new acid mine drainage treatment plant to treat 4.2 billion litres of contaminated runoff per year, removing 226,000 kilograms of heavy metal contaminants. The amount of copper removed yearly is equivalent to preventing thirty million copper pennies from entering Howe Sound. The heavy metal pollution was reduced by 99 percent.

After a long period of abuse and neglect, the waters of Howe Sound were coming alive again. These expensive efforts, combined with the presence of the glaciers and the rainforests, restored the water quality of Howe Sound. This is proof, says Mark Angelo, that "we should never give up on a river."

The first sign of recovery was the return of small marine life—forage fish, such as herring, surf smelt and sand lance, some of the least studied fish in the sea. There is no legislation to protect their habitat, yet they are essential to the health of our oceans. Not the kind of fish you're likely to see on a poster, nor listed in Canada's Food Guide, forage fish are tasty and they used to be plentiful. They are the foundation of the food web on the west coast, a major source of food for seabirds, salmon, ling cod and rockfish, which in turn, feed whales, sea mammals, and humans. For the first time in many years, the table was set for the bigger fish. Crabs and prawns were also being caught throughout the Sound, and commercial fishing was once again allowed in the Squamish River.

In 2015, I jumped at the chance to do some citizen science research on a Bowen Island beach with Ramona de Graaf, a marine biologist who has worked with over thirty communities in BC, training them to find forage fish embryos on local beaches. The coast of BC once had millions of forage fish; today, in spite of improvements, the populations are still low. As a result, many of the birds, fish and mammals that rely upon them are also suffering. Ramona is determined to turn that around. "It's more than just a pretty beach," says Ramona, as several of us in rubber boots follow the beacon of her red hair and her siren song down to the beach. "That's where the forage fish spawn."

Not that long ago, so one of the old-timers tells me, you could go to Sandy Beach and easily gather a bucket full of delicious smelt for supper. That's where we're headed on a clear bright morning in search of embryos. For me, the beach is no longer just a place to swim, sunbathe and make sandcastles; it's a nursery. "Fifty percent of the diet of the chinook salmon is Pacific sand lance," Ramona says. If you want salmon, you need forage fish. And if you want orcas you need chinook salmon, the main diet of the iconic Howe Sound whale.

We are motivated. We measure off thirty metres in between the high tide and the low tide where they are likely to spawn. Ramona instructs us to have faith even if we don't understand all the data and the science. "Fish don't spawn according to the literature," she cheerfully reminds us. We get to work. We shovel gravel. We sieve. We add more sea water. We sieve. We sing. We winnow—much like a belly dance as you shake and stir the pan.

Ramona compares it to panning for gold, but the treasure we seek is the smelt embryo.

Ramona, known as Queen of the Smelts, tells the story of a typical surf smelt beach romance. The female surf smelt swims onto the shore, and when she feels the beach's pea-sized pebbles on her body, she lays her eggs in the sand; the male, following closely behind, lays his sperm on top of the eggs. Once an egg is fertilized, it forms a suction cup that attaches to a pebble for safety. After that, the "kids" are on their own. We all agree to return to the same beach every few weeks to take samples, winnow, winnow, winnow, gather data and send them off to Ramona. "Bowen is in a good position," Ramona tells us. "Your neighbour, West Vancouver, has plenty of surf smelt caught commercially." To date, we have yet to find the much sought-after surf smelt embryos. However, the courtship of the anchovies has been recently spotted in several locations on Bowen Island and elsewhere in Howe Sound.

There's a lot of romance going on in my waters. That's what makes my world go around. Not too many people writing novels about it, but it's something to behold. Especially the whales. The water might be chilly but it does not affect their orca ardour. There is a long courtship. There's kissing. There's dancing. A slow and passionate water ballet. There's cuddling. Lots of cuddling. They're mammals after all.

There is no doubt plenty of romance going on in the Squamish River Estuary, and it's happening right near the centre of town. The estuary may not have the immediate dramatic appearance of the mountains or waterfalls of Howe Sound, but it's one of the most productive natural habitats in the world, and has made life here possible for thousands of years. When salt water and fresh water come together in an estuary, they create an idyllic environment for the enhancement of life, both human and other than human. Estuaries also filter the water and act as nurseries for small creatures.

There are two other estuaries in Howe Sound—McNab Creek and Furry Creek. Estuaries provide perfect accommodation for many species and are essential for the survival of plants and animals, birds and amphibians, fish and many other aquatic species. This is critical habitat when one realizes that estuaries in British Columbia cover less than 3 percent of the coast while providing habitat that is used by 80 percent of the species found on the coast.

Edith Tobe has generously agreed to take me on a tour of the Squamish River Estuary. Edith is a habitat biologist and executive director of the Squamish River Watershed Society. I am immediately attracted to her earthy manner and hearty laugh. She has worked in the resource sector as a resource engineer technologist. While working with logging companies, she began looking into assessments on watercourses in the Squamish River for the possibilities of restoration. She studied erosion, nutrient deficiency, barriers to fish and the presence or absence of fish. "I'm familiar with every watercourse around here," she says. "I've floated, swum and boated here." She tells me the Squamish Estuary, the ecological jewel of Howe Sound, was the reason she decided to move to Squamish twenty-three years ago. I am about to find out why.

Edith leads me on a leisurely walk in a green oasis to the accompaniment of birdsong and the silent rustling of leaves. This wasn't always the case. In the 1970s, the Squamish River Estuary was cut off from its fresh-water source when BC Rail developed the Squamish River training dike. The idea was to move the river toward its west bank and dry out the central estuary for the development of a proposed deep-sea coal port. The Department of Fisheries and Oceans eventually shut down the project but the dike remained, and the large piles of dredge material were never removed. In 1979, DFO struck the Squamish Estuary Committee in partnership with the BC Ministry of Environment and community stakeholders to create a vision for the Squamish waterfront.

For nearly thirty years, the central estuary at the mouth of the Squamish River lay buried under thousands of yards of river dredge material. It looked more like a moonscape than an estuary. In 1993, Edith helped form the Squamish River Watershed Society, and they became a non-profit in 1998. The following year, they started work restoring the former dredge spoil area by constructing culverts across the dike and then removing the dredge spoil material. It took the work of countless community volunteers in combination with BC Rail and the DFO to restore the estuary.

One of the first restoration channels was constructed in 1999. Today, it has the look and feel of a completely natural, functioning ecosystem. It's hard to imagine that it hasn't always been there. The channels were built by putting culverts through the dikes. "Now you can kayak at high tide. But we had to disturb the whole area to build this channel. A big excavator dug down, casting the soil on one side to make a steep slope, keeping the other side shallow for ducks to wade in and out." After they built the channel, they organized planting parties to replant all native species—sweet gale, salmonberry. "First you dig it, then you scrape it and the water fills and the plants just come. Six months later it's a natural habitat."

Edith started working on eelgrass restoration in 2001. "We wanted to create habitat for herring. I knew we had eelgrass historically. Colin Levings, then with Fisheries and Oceans, had seen eelgrass in the Cheakamus River, a tributary of the Squamish River. I could see some of it going by. I met up with a grad student and we got some eelgrass plants from Bowen Island and the Sunshine Coast to see if it was feasible to reintroduce eelgrass."

Edith's devotion to the estuary is inspiring. She talks about eelgrass, which she calls "Meadows of the Sea," with all the excitement and anticipation of a gardener contemplating the planting of a rare species. The eelgrass is now sheltering herring and many other fishy things, including crabs, who lie low until their shells harden. Eighty percent of fish species and marine invertebrates use eelgrass during some part of their life cycle. It is also a resting and feeding habitat for migrating birds.

We enter a green, inviting channel of the estuary that was once devastated by logging and other industrial activity. "A desert," she says. "No crabs, no clams." Then they put in the eelgrass plugs. "Twenty-four hours later, there were crabs, fish." Just at that moment, a great blue heron, startled by our presence in her home, flies up, her curved neck and long elegant body, her massive wings moving with extraordinary grace through the air. On her way to feed on herring perhaps.

I ask Edith where herring like to spawn if they have their druthers. Seems they're not particularly discerning. "If you put your hand in the water," she says, "herring will spawn on it. But their survival rate is highest in eelgrass beds, which provide shelter and move with the tides, keeping the herring eggs under water." When the eelgrass beds in Squamish were eradicated by development, the ever-determined and adaptable herring began to lay their eggs on the wood pilings at the Squamish docks. The pilings, coated in creosote, resulted in 100 percent mortality for the herring eggs.

Creosote is another good news story gone bad. The idea behind creosote was that it would stop the wood from rotting and more trees would be spared. This would be better for the environment, and better for the company bottom line. "The problem is creosote is toxic to the water and anything that gets near it," says Edith. The answer "is to pull those pilings out, but the government has no legislation or policy discouraging the use of creosote."

In 2006, the Squamish Streamkeepers, a volunteer group, began wrapping the pilings with protective cloth. This effort, along with the cleanup of toxins in Howe Sound, the restoration of the Squamish Estuary, the closing down of the Woodfibre Pulp Mill, and the presence of eelgrass, has led to a return of the herring, an important forage fish. Today, the estuary is habitat for more than two hundred species of wintering and migrating birds, as well as bear, muskrat, deer, otter, great blue heron, peregrine falcon, cutthroat trout and five species of salmon. It's an inviting natural respite close to downtown Squamish.

In 1993, Edith began a process to bring back the salmon, who need to spend time in the estuary as part of their life cycle. In the 1970s, they were cut off from getting into the estuary. By 1972, pink salmon stocks had disappeared. Today, the slow-moving tidal channels lined in eelgrass and sedges provide space for juvenile salmon to feed, grow and hide from predators. The roots of the cattails, shrubs and trees in the estuary filter and purify the water before it enters into the Sound, and soil in the estuary prevents floods by absorbing excess water in the rainy season.

The estuary continues to be stewarded and monitored by the Squamish River Watershed Society and other community groups. For Edith, it's more than just a nice place to go for a walk, to restore the human spirit after a long day's work. "There's a big push for nature as recreation but very little for nature itself," she says. "I build habitat for nature."

The estuary is such a gift for every creature that swims or flies or walks. It's gone through plenty of hard times since the first Newcomers came. This new breed is doing things in a good way. They know the best way to thrive is to be in partnership with Mother Earth. To accept her teachings, her limits. To recognize this land, these waters as powerful medicines. The Newcomers don't really understand the notion of medicine, though. They think it's something you buy at the drugstore. For the First People, medicine is seen as energy, power, spirit, relationship. It is a movement, a balance, a way of life.

As we leave the estuary, I ask Edith about the industrial complex that shares the estuary. "That's the Squamish Terminals," she says, "an important industry in Squamish. It provides much needed employment in the resource sector and allows for the shipping of wood fibre." Squamish Terminals is the only deep-sea port in

Squamish and connects directly to rail and road access, allowing for the movement of a large amount of material. "As with any industrial port, there is always the risk of contamination and pollution," she says.

Edith is referring to an environmental incident that occurred in August 2006. "A ship—the *Westwood Annette*—breached her hull and spewed Bunker C oil into the Squamish Estuary. The residue of the oil remains in the sediments adjacent to the spill site, where it will take tens of thousands of years to break down. Due to the location of the Squamish Terminals within the Squamish Estuary, there is also the conflict of the impacts this type of industry has on the overall estuarine function. Compounded by the rail line that services the Squamish Terminals—the Spur Line, managed by CN Rail, which bisects the Squamish Estuary—the impacts are quite significant."

Edith, who has worked in the resource sector, is always concerned about the balance that exists between industry and the environment. "When the Squamish Terminals was constructed in the 1970s, there was a lot less awareness of the importance of estuaries and the potential impacts heavy industry can have on the environment. In today's world, we recognize we still are quite ignorant of the total impacts humans have on the environment, but we are hopefully taking a more precautionary approach so we will minimize our contamination of the essential elements of life, including clean air and water."

What about the future of industry in Howe Sound? "Hopefully as we move forward," says Edith, "we will see a reduction in industrial and other developments within our planet's most sensitive habitats. But for those industries that currently do exist, it is important that we always demand the highest standards in their operations to ensure minimal pollution and risks to nature."

I want to find out more about Squamish Terminals and discover how it stacks up as a corporate citizen. I make an appointment to speak with Kim Stegeman, President, and Doug Hackett, Director of Special Projects and Information Services. After I go through security—beefed up after 9/11, they tell me—I am warmly welcomed by Kim and Doug. Squamish Terminals is a busy place, the wholly owned subsidiary of a Norwegian company. The first ship arrived in November 1972. Between sixty to eighty ships are received or sent out every year, supporting many communities in western Canada. One hundred full-time employees, the majority of whom live in Squamish, load and unload the ships, rail cars and trucks.

Much of their cargo is wood pulp from many parts of BC on its way to Asia. They have three warehouses and two berths. The east dock was rebuilt after a fire in early April 2015. "While the fire was devastating and affected our operations," says Kim, "it did provide an opportunity to build a more modern structure." During the rebuild, the company retained all their employees. "We wanted to give our employees certainty," says Doug, "and show the community we have a long-term commitment. As the ashes settled, the terminal got busier, so the decision to keep everyone on turned out to be the best business decision as well." For the future, they are looking at getting more steel from Asia or Europe for construction in the Lower Mainland, as well as for pipelines and rail tracks. Not surprisingly, they welcome the proposed LNG plant.

Squamish Terminals has become an integral part of the community, a strong supporter of local business as well as community organizations and events. Doug is chair of the Squamish Community Foundation. A self-described right-wing industrialist, he tells me he likes solutions. Solutions are essential when your business operates in an environmentally sensitive area—the Squamish Estuary, as well as the Squamish Harbour, site of the annual herring spawn and habitat for salmon fry. The Terminal has been working with the Squamish Streamkeepers since 2006 to enhance herring habitat under its docks.

In April 2012, the company installed four new floating fish pens under the dock with the aim of increasing the survival rate of salmon smolts hatched and raised at the Tenderfoot Hatchery. The smolts are placed in the pens, allowing them to become accustomed to ocean conditions while being protected from the fjord's winds and marine animals. The Terminals have also replaced the burned creosote pilings with pilings made of steel, and they ceased construction of the new dock for two weeks during the herring spawn.

Kim has a background in the hotel business. I ask her if she is bringing hospitality to the shipping business. She tells me about the "Great Christmas Tree Experiment" in 2014, a partnership with the Streamkeepers. The idea was to collect abandoned Christmas trees—the bringers of joy over the winter holidays—and see if they might also bring habitat joy to the herring. The Boy Scouts collected the trees and placed them in the Squamish Harbour in time for the herring spawn. The experiment, a nice community-building exercise, was, however, not particularly successful. "They liked the derrick better," Doug laughs. "Less wave action, more stable."

There are many others keeping watch over the fish in Howe Sound. The Bowen Island Fish and Wildlife Club, a volunteer society formed in 1967, has been involved in the federal government's Salmonid Enhancement Program (SEP) since 1982. They also continue to work at stream rehabilitation, and promote the protection of rockfish in and around Bowen. Every year, under the supervision of Fisheries and Oceans Canada, between 250,000 and 400,000 pink, chum and coho salmon are incubated and raised in the Terminal Creek Hatchery.

The Coho Bon Voyage takes place in June every year. This event gives young children and their parents an opportunity to learn about the fascinating life cycle of salmon and to participate in their release. It's one of my favourite community events on Bowen, watching toddlers as they carry their small white pails full of squiggly, healthy coho fry to the creek, and gently tip the pails to set the fish on their journey. Goodbye! I've heard the children say. Have a good swim. See you in four years. I believe in you!

Over the last ten years, spawner returns to Bowen have been disappointing. So there was great excitement on the evening of October 19, 2016, when twenty chum—these are the big ones—were spotted as they entered the fresh-water lagoon. Tim Pardee, often referred to on Bowen as "the salmon guy" because of his passion for wild salmon, and other SEP volunteers began to monitor spawners and get the word out to the community. News spread quickly on the island, and the following morning, a crowd gathered to greet the chum as they swam from Deep Bay into the lagoon. While some spawned right on the gravel bed at the entrance to the lagoon, others headed up to the top of the lagoon and reached the bottom of Bridal Veil Falls. Another fifteen were counted. We were in luck. One pair very graciously performed their mating and spawning ritual right in front of us. Everyone was thrilled, including the harbour seal in the bay, bobbing its head up out of the water and licking its lips.

For over four weeks, the run continued, and the lagoon became the favourite island gathering place, after the library and the pub. One Sunday, about thirty of us watched as many salmon tried to negotiate the ramp to get into the lagoon. The tide was not quite high enough for easy entrance. As they struggled to wiggle themselves up, a cheering section was created. "Come on, you can do it! Yes! Yes!" followed by disappointed groans as the weary salmon slid back into the bay. A gentleman from Ottawa told me that watching salmon spawning was "the thrill of a lifetime." He was honoured to witness the end of this amazing life cycle. I suggested that on his return to Ottawa, he might pay a visit to "those people on the Hill," let them know how important salmon are to the west coast so they might make well-informed decisions.

Children also got in on the act. One day, there were eight chum who got stuck on the rocks near the lagoon. There was little chance they would survive until the tide rose. Three ten-year-old boys decided to do something about it. They climbed down onto the slippery rocks and heroically lifted each salmon—they look even bigger out of the water—and carried them carefully to safety. Over twelve hundred chum and coho spawners returned to Bowen Island in the fall of 2016. For many, the experience at the lagoon was an emotional one, something they will never forget. Observing the return of the salmon gave us the feeling that we were a part of something much greater than ourselves, and that we were finding a way to live in harmony with it.

American scientist and environmentalist Aldo Leopold, who died in 1948, had this to say in *A Sand County Almanac* about how one should live in harmony with the bioregion. "A thing is right when it tends to preserve the integrity, stability, and beauty of the biotic community. It is wrong when it tends otherwise." His practical wisdom could encourage us to widen our sense of community to include watersheds, plants, animals and soil, and to offer them a worth beyond their immediate value to humans. In doing so, humans could see themselves as part of the Earth community, in which we are all embedded. This Earth community is the source of all creativity and sustenance. Once we apply for membership in a biotic community, we cease to be its exploiter.

I get the chance to meet Randall Lewis once again. President of the Squamish River Watershed Society since its inception in 1998, Randall is passionate about the society's restoration of the Squamish watershed. He invites me to observe the remarkable improvements to Evans Creek, begun in 2013. "The society has a list of priorities for which waterways to restore," he tells me, "and Evans Creek has always been at the bottom of the list. But it's one that's very important to me." Evans Creek is near Randall's home. It was a historic river with ancient fish weir sites identified by the Elders. There was evidence of a historic channel that had been cut off by dikes built to protect hydro and residences. These dikes diverted water and cut off salmon's access to the channels.

"Because of the knowledge of the Elders, we knew the channel was here." With great pride, he takes me on a tour of the revitalized Evans Creek. "What we're doing here is reconnecting the arteries of the land, bringing life back to the land." It's a typical west coast day, on-and-off rain, but mild. The land welcomes us with the soothing sound of running water. There is not another human in the area. I inhale the silence and the beauty. Randall has not forgotten what it means to belong to a place, to be willing to protect the land and the waters that have given his people their very life for thousands of years.

We approach the Moody Channel, where the society is creating salmon habitat by placing four intakes on the dike and building artificial log jams to keep back the gravel. Salmon spawn on gravel. "Different gravel for different species," says Randall. "Chum and coho spawn anywhere, while pinks and chinook are more fussy. And when the Fraser River gets too hot for the sockeye, like swimming in a jacuzzi, they stray to the Cheakamus River or the Mamquam. We're providing a five-star hotel for spawn and fry where they're safe from predation." As each channel is opened and the waters run free, Randall posts videos on Facebook showing the return of chum and pink salmon to the waters that nourished and sheltered them for thousands of years. In November 2016, two coho salmon were spotted in Evans Creek. "Build it and they will come." It's a remarkable story of resilience, persistence and deep caring for the land.

The recovery of these waterways is beneficial to more than just salmon. "The salmon feed the land too," he says. "Fish carcasses get distributed through the forest by bears, and eagles fly carcasses to the treetops, providing nutrients for the trees, creating a rebirth, a regeneration of the area." Children in Squamish schools are also involved in the restoration project. Grades one to seven help with the rearing and releasing of the pinks. "That's my fish, they can say four years later, when the pinks return. They have ownership."

The next phase of the restoration will be an educational walkway through the creek for Elders and youngsters to rest in the beauty here, to listen to the land, and to appreciate the life cycle of salmon. Since 1992, many miles of channel have been restored. "This is a model," Lewis says, "a success story. Today we're hiring biologists for the work. One day soon we want our own youth to become biologists."

The renewal of the waterways in Howe Sound is reflected in important cultural renewals for the Squamish Nation, including the revival of the seagoing canoe. The Europeans did not discover Canada. They were shown it. By canoe. As someone who has enjoyed kayaking in Atl'kitsem/Howe Sound, I'm intrigued by the canoe culture of the west coast. I first heard about the Tribal Canoe Journeys from Roy "Bucky" Baker, a paddler on the Squamish Family Seagoing Canoe. Bucky is a Squamish language teacher, as well as a member of the Squamish Nation Band Council. He holds a deep knowledge of Squamish culture and tradition, and somehow finds time to make the most stylish cedar hats, including an accurate and astonishing Darth Vader version.

I have attended several of his ceremonies and always come away knowing that I have been witness to how ancient traditions continue to resonate in our contemporary world. Bucky tells me the canoe journeys have been a powerful experience for him. He describes how "the Elders saw their youth drifting away from their culture, getting lost in drugs and alcohol. They decided to teach them their traditions through the canoe." Bucky suggests I contact Bob Baker, who was instrumental in reviving the seagoing canoe and the canoe journeys.

Bob, whose ancestral name is S7aplek, is an Elder with the Squamish Nation. He is steersman for Tribal Journeys and Pulling Together Journeys. He is also cultural consultant to School District #45. I meet Bob in a coffee shop in West Vancouver's Park Royal Shopping Centre, smack dab in Squamish Nation territory. Bob is friendly and eager to talk about his life and the revival of the great canoes. We sit by the window and our delightful conversation is pleasantly interrupted by passers-by. Bob catches someone's eye and smiles.

"There's my friend Evan Adams."

"Hmm, he looks familiar to me."

"Yeah, he was in *Smoke Signals*."

"Right! I loved that movie."

Smoke Signals was a groundbreaking movie, written by Indigenous author and filmmaker Sherman Alexie, based on his book

The Lone Ranger and Tonto Fistfight in Heaven. It was hilarious and moving, and starred a stellar cast of Indigenous actors. Nothing like a little star power to set the mood for a good conversation.

We have our coffees, a "Tall Pike" for Bob—in honour of the years he lived in Seattle, and a decaf Americano for me—in honour of sparing my guest the havoc that caffeine inflicts on me. Now I can get to my questions. Bob was raised on Squamish Nation territory and lived in Hawaii during the late 1970s. A beach bum, he tells me, "first and foremost a surfer and beach bum, I carried my surfboard wherever I went." He worked as a rigger on ships in Pearl Harbor and was head of maintenance in a Waikiki Hotel. He has always been passionate about the canoe and owns seven of them, including one made by the Ainu in Japan.

Bob says it was Frank Brown, a Heiltsuk Nation member, who started the revival. The theme for Vancouver's Expo 86 was transportation, and Frank wanted to show the world the earliest form of transportation on the west coast: a traditional seagoing canoe. "He made himself a canoe and paddled to Expo," says Bob. "He was making a cultural statement." Three years later, Frank went to Seattle for the first canoe journey—Paddle to Seattle— part of the one hundredth anniversary of Washington Statehood. Fifteen nations participated, stopping at villages along the ancestral Salish coasts of British Columbia and Washington State.

Bob heard about the Paddle to Seattle and wanted to involve Squamish Nation in the next journey to Bella Bella. "In 1990, I approached chiefs and council about getting a canoe for our nation. I was lobbying, talking about the canoe movement," he says with emotion in his voice. He wanted to make a seagoing canoe. The problem was nobody knew what they were. "You can see what my task was. I spoke to the Elders. I knew we had to have big canoes to go travelling to Vancouver Island and the Inside Passage. Archaeologists found we were trading with people in Oregon. We needed those big canoes to freight things." But the knowledge of the making of those canoes had all but disappeared.

He went to museums in Washington State to begin his research. He found a professor at the University of Washington who had the plans for a seagoing canoe. Then he went to Tulalip to meet with carvers and documented everything. "I took a cameraman with me to film it. The carver showed us his canoes while we ate a big pot of salmon stew. There were one-hundred-year-old canoes in a chief's shed. Two of them. They were considered sacred so we couldn't take any photos."

Bob spoke with many carvers, several of whom were willing to make the canoe, but they wanted to build it in their own way. Bob wanted an exact replica of what his ancestors used. He knew Cedric Billy, a member of the Squamish Nation, was an expert canoe maker. Although Cedric had made seventy canoes, he had never steamed a canoe, the process needed to make a seagoing canoe. But when Cedric said he would make it the way Bob wanted, Bob gave him the plans and he began carving in 1991.

Seagoing canoes are crafted from a single log, which can be hundreds of years old. For this canoe, they got a log from the Sto:lo Nation. It takes about a year to carve the big canoes. "Racing canoes are just carved on the go," Bob tells me. "They're made in a hurry. Bigger canoes need to be steamed so they will be more resilient." Once the log has been cut, it is left to season over the winter. In the spring, the carver hollows out the interior. It is then ready for steaming. Bob describes how the canoe is filled with water, and then hot rocks are placed in the water. As the water heats and the steam rises, the wood becomes pliable and the canoe is stretched. This causes "the belly to drop and the sides to open, giving the beautiful lines." One of the special things about the ocean-going canoe, Bob tells me, are the grooves underneath the hull—approximately two centimetres wide and one-third of a centimetre deep. "This is to agitate the water so there is no suction to the hull; the canoe travels on bubbles. There is no drag, so the canoe is very maneuverable."

At this moment, a beautiful young woman walks by the window. "That's one of my dancers," says Bob, who co-founded Spakwus Slulum, a dance group I have had the pleasure of seeing at various ceremonial events. Bob tells me the group's first invitation was to perform at the Montreux Jazz Festival in Switzerland in 1993. "Start from the top and work your way down," he laughs. After getting her coffee, the woman comes over to our table and Bob introduces me to Devon Williams. I discover she is the daughter of Linda Williams, whom I've met on several occasions. After a brief heart-to-heart exchange with Devon, it's back to our canoe talk.

While the canoe was being built, Bob threw down the gauntlet to the Squamish Nation to paddle to Bella Bella. "I put out a newsletter to let people know I was looking for pullers—our word for paddlers. After six months, I wasn't getting much feedback, so I had to hand-pick my pullers." He needed a Squamish language speaker, so he chose his nephew, Chief Ian Campbell. He also needed a spiritual person who would help with preparations and conduct sweat lodges. That was his cousin, Richard Baker. As well as experienced paddlers like Gordon Newman, he needed artists to provide gifts for the chiefs, Elders and youth. "That's three gifts at every stop along the way." He chose Rick Harry (Xwalacktun) and Aaron Moody, accomplished Squamish carvers, and Klatle Bhi (pronounced Cloth-Bay), a Kwakwaka'wakw and Squamish artist, who provided prints for each landing. "It was like pulling teeth to get the first group," he says. "Then after the others saw what we were doing, they wanted to be part of it."

Depending on distance, the canoe journey can take up to a month. It's hard work being on the water eight to ten hours a day. "And you never leave a canoe alone," says Bob. "It's a living member of the family. A tree is alive until it's no longer there." It took them three weeks to get to Bella Bella for the great canoe gathering, "but we could have done it in ten days." The stops along the way slowed them down. Protocol must be observed: upon arrival, the canoe skip introduces the members of the crew prior to asking permission to come on the land. Each canoe is welcomed individually by a representative of the host nation and invited to come ashore. The sharing of songs and dances and gifts can last for days.

"We arrived in Sliammon, near Powell River, around nine o'clock at night. They were so respectful. They lifted our forty-five-foot canoe with fourteen paddlers in it and placed us on smooth logs. Next morning, when we asked permission to leave their territory, we were refused." The Sliammon urged them to stay on another day. They had lost all information about their canoes, and wanted the Squamish paddlers to share their knowledge of the wonderful seagoing canoe.

The journeys continue to grow in popularity. Thirty-nine canoe teams participated in the Bella Bella journey. In 2016, 120 canoes journeyed to Nisqually at the south end of Puget Sound. There are two journeys every year: Pulling Together, a journey with the police, and the Tribal Journey. They involve more than paddling. The journeys enable the passing down of ancestral teachings and traditions, including ceremony, especially to youth. The journeys are drug- and alcohol-free, Bob says, a time for spiritual and physical transformation. For the Squamish Nation, the journeys have become part of a cultural and spiritual awakening.

I've often enjoyed sitting in a canoe or kayak, listening to the sound the paddle makes as it strokes the sea. The air is sweet and salty. Sometimes when the wind is benevolent and blowing from behind, you hold up your paddle, catch the tailwind and move along very nicely. Other times the howling wind stops you cold, rearranges your face. I've had some challenging moments in Atl'kitsem when the waves were deep troughs, and it seemed as if the kayak was about to be swallowed by the sea. It gave me a tremendous respect for the sea. I learned to travel with the waters, not against them. And what I've learned from being on the water has helped me when I'm back on land.

The land has always been an integral part of the spiritual and economic life of Squamish Nation. On a visit to Totem Hall in Squamish, I chat with Joyce Williams again. Joyce is a cultural teacher at several schools in the Squamish area. Every year, the students choose a topic to research. This year they chose the environment, with a focus on Woodfibre LNG, the proposed liquified natural gas plant near Squamish. "They all had different concerns: herring, greenhouse gas emissions, ocean health." One of the biggest problems, she says, is that "the young people are losing that connection to the land."

She was thrilled to be part of a group that brought several youth on an overnight camping trip to Kw'em Kwem, also known as Defence Island, historically very important to her nation. "Since time immemorial," she tells me, "the Squamish people used and occupied the island and it is well-known as a Squamish burial site, as well as a fishing station." The response from the youth was gratifying. "One of my biggest hopes is to bring them out on the land. There they find out what's taking place in their territory and they become engaged."

Linda Williams, Joyce's mother, is on the board of the Squamish Nation Parent Advisory Council, working closely with the Education Department to assist students with funding for sports, recreation, education and hot lunches. When Linda was a young person, it was her job in the family of fourteen to keep the fire going. She did this along with her brother Randall taking shifts. They both continue to keep the fires of their culture going.

"Over the past seven years," says Linda, "we finally got a really good working relationship with the school system, everyone sitting at the same table, with superintendents and principals. We worked hard with the principals to get programs like Cultural Journeys for children in K to six, and Learning Expeditions for grade six to high school."

These programs are part of an Aboriginal Education Enhancement Agreement, a five-year program implemented in 2014, a commitment by school districts, local Aboriginal communities, and the Ministry of Education to work together to support Aboriginal learners. There are homework and culture clubs for students, and the teaching of the Skwxwú7mesh sníchim— Squamish language—is under development. Linda's face lights up. "Through dance, through drumming, through language, our young people are getting reconnected to their culture."

It's estimated that 50 to 90 percent of the world's six thousand languages will disappear in this century. The Skwxwú7mesh sníchim is one of thirty-two languages and fifty-nine dialects spoken in BC before the arrival of the Europeans. Seven years ago, the language was on the brink of extinction, with fewer than ten fluent speakers. Khelsilem Rivers, a community organizer and graphic designer, is committed to doing something about this. Born on the reserve in North Vancouver, he was given the traditional name Khelsilem by his paternal grandmother.

In January 2016, he launched Kwi Awt Stelmexw, a new initiative to save the Squamish language. The not-for-profit organization partnered with Simon Fraser University to launch a full-time language immersion program. Khelsilem began teaching the language five years ago. "I was giving volunteer language nights in my father's home with only a handful of people. And these grew to ten, twenty, and more coming on a regular basis. There is a growing hunger among our people for their language." By the end of the current school year, the fourteen students enrolled will be fluent enough to teach others.

Language gives expression to the land that birthed it. The Squamish language emerged from the watery heart and mountainous bones of Howe Sound, from the breath of the Squamish wind, the heat of the volcano, the movement of the rivers and the people. The reclaiming of the Squamish language and the stories embedded in it are an essential element of the renewal of Howe Sound. These are the sounds that resonated for thousands of years throughout the land; the land longs to hear them again.

In *Love of Country: A Journey Through the Hebrides*, Madeleine Bunting takes a thrilling voyage through the remote islands of the Hebrides and considers the effect of language on identity and culture. For example, the Gaelic words for land make it clear that land is not owned; land is shared, and bound up with relationships to others and to responsibilities. "People belong to places, rather than places belonging to people."

Because of the strength of their language, the people of the Hebrides have been able to protect their land from outside developers. Bunting cites a case where the developer used "maps that removed all the Gaelic names and replaced them with numbers; words such as 'wild,' 'remote' and 'empty' were used in a deeply political way to suggest that the land had no value." Capitalism requires "a place-lessness," Bunting writes, "to ensure the smooth flow of capital, people and resources to achieve economic efficiency." She discovered that the Gaelic language continues to provide resistance to these ideas.

Anthropologist Wade Davis talks about language as "an old growth forest of the mind, a watershed of thought, an ecosystem of social and spiritual possibility." But for most westerners, language is generally thought to be simply a medium in which to express our thoughts and feelings. "We tend to think the languages we speak are like different radio sets," says physicist David Peat in his book *Blackfoot Physics*, "the passive vehicles for our ideas. Within Indigenous science, however, language has a power all its own, and to speak it is to enter into an alliance with the vibrations of the universe."

I'm intrigued by the idea of entering into alliance with the vibrations of the universe through the languages I speak and the songs I sing. It seems clear to me that people can no more live without their language than a tree can grow without roots. I am very proud of my first language, French, and enjoy speaking it whenever I get the chance. So I bought myself a Squamish-English Dictionary to enter into that particular vibration. I'm learning that the Squamish language contains the stories of the land, stories of the great journeys that took place on the waters, and the spiritual alliances that go back thousands of years. I was a bit startled to see the number 7 in a word until I found out it represents what is called a glottal stop. A bit like the sound you make when you see something amiss. "Oh! Oh!" Take the word, ha7lh. It means "good." So you make that glottal stop in the middle of it. Ha (glottal stop) th. More or less.

I learn new expressions and try them out when I'm with a Squamish speaker. In conversation with Roy "Bucky" Baker, I got brave and tried out my few phrases. "Ha7lh Skwayel! (Good day!) I even tried pronouncing the original name of Bowen Island: Kwilákm. "How was that, Bucky?" "Close enough," he answered. Then he smiled and said, "When you speak our language, it makes us strong."

At a conference in Vancouver in May 2017, where writers were gathered to discuss how we might respond to the Truth and Reconciliation Commission Report, writer Julie Salverson helped me see the blessing in Bucky's words. "It was a benediction," she told me. "Close enough. You don't need to be perfect to carry on the work of reconciliation."

Several years ago, I was honoured to take part in the Truth and Reconciliation process in a weekend workshop with Chief Robert Joseph. The TRC was a healing balm that moved across the country as we acknowledged Canada's brutally destructive relationship with Indigenous people. A good beginning, and we still have a long way to go. Reconciliation between Indigenous and non-Indigenous, and restitution for stolen lands, is crucial to the continuing health and prosperity of Canada.

One of the recommendations of the TRC was to "repudiate concepts to justify European sovereignty over Indigenous lands and peoples, such as the Doctrine of Discovery. In 1493, a papal bull from the Vatican declared that when European nations "discovered" non-European land, they gained special rights over the land, such as sovereignty and title, regardless of what other peoples lived on that land. The tragic results of this doctrine have continued over five hundred years. One aspect of reconciliation that is rarely mentioned by non-Indigenous leaders is that true reconciliation must include reconciliation with the Earth. And it must happen very soon here in Howe Sound if the renewal of the waters and the return of marine life is to continue.

In February 2017, the Vancouver Aquarium's Coastal Ocean Research Institute (CORI) published its report on the health of Howe Sound, "Ocean Watch: Howe Sound Edition." The launch was attended by Fisheries and Oceans Canada Minister Dominic

LeBlanc, Howe Sound politicians, contributors, media and VIPs, including Consul General LIU Fei of the People's Republic of China. I'm indulging in a bit of name-dropping here to point out the significance of this report. It's an in-depth study of the Howe Sound ecosystem, bringing together years of research, touching on "ecological, socioeconomic, cultural and governance aspects of ecosystem health." Its stated intent: "to provide information to help guide decisions as the area grows and changes."

The report provides a big-picture look at Howe Sound and integrates extensive knowledge from First Nations, governments, academics, NGOs, citizen science and industry. "Two decades of revitalization efforts have returned large areas of the Squamish Estuary to a vital wildlife habitat," says the report, "and reversed the effects of human activity and industry. Yet, the impact of industry and human intervention will be felt for a long time."

The data are startling, a wake-up call. The renewal that we are observing—the return of life to our waters—is in a very vulnerable state. Less than 1 percent of Howe Sound is protected under provincial legislation, and nearly half of the species and habitat researched are rated critical according to the CORI criteria, while the other half either lack data, require restoration or remain unprotected. Salmon, rockfish and ling cod are among the species classified as critical. Howe Sound remains vulnerable to development, shipping and fishing.

In my interviews with locals and visitors, there was a lot of enthusiasm and hope for the continued renewal of Howe Sound. So many are enjoying the return of species that had been absent for decades. But their enthusiasm could be short-lived if governments are not committed to a full recovery. This would mean toning down the push to industrialize, allowing the waters to recover, and making peace with the Earth.

⏤

Reconciliation is about relationship, about living together with respect. Respect for people, respect for the land. When you respect the land and the waters, you get such a tremendous tailwind from the ancestors. You move better and smarter. You make good decisions. Doesn't matter what you call her—Nature, the Environment, the Planet, Pachamama. It's all the same. Mother Earth is what nourishes your existence every moment of your life.

꿏

The acclaimed author and conservationist Roderick Haig-Brown (1908–76) said there were two kinds of citizens in North America: "the Boomers and the Stickers." The Boomers were "those who pillage and run and want to make a killing and end up on Easy Street." The Stickers want "to settle and love the life they have made and the place they have made it in." Fortunately, most of the people I have met on this Howe Sound journey are Stickers.

I find inspiration in the words of Robin Kimmerer, Professor of Environmental and Forest Biology at the State University of New York College of Environmental Science and Forestry. She draws on both Indigenous knowledge—her grandfather was a member of the Potawatami Nation—and her scientific knowledge as a plant ecologist. In her TED talk, "Reclaiming the Honorable Harvest," she presents a model based on Indigenous wisdom for living well on the land. "Demand an economy that is aligned with life," she says, "not stacked against it. Becoming Indigenous to a place means living as if your children's future mattered, to take care of the land as if our lives, both material and spiritual, depended on it."

The honourable harvest demands that we ask permission of the land before taking anything, that we listen for the answer, that we take only what we need and waste none of it, that we minimize harm and share the harvest. Most of all that we acknowledge our gratitude. Social, political and economic decisions based on the honourable harvest could lead to an enhanced future for generations to come. "We have chosen the policies we live by," Kimmerer says. "We can choose again."

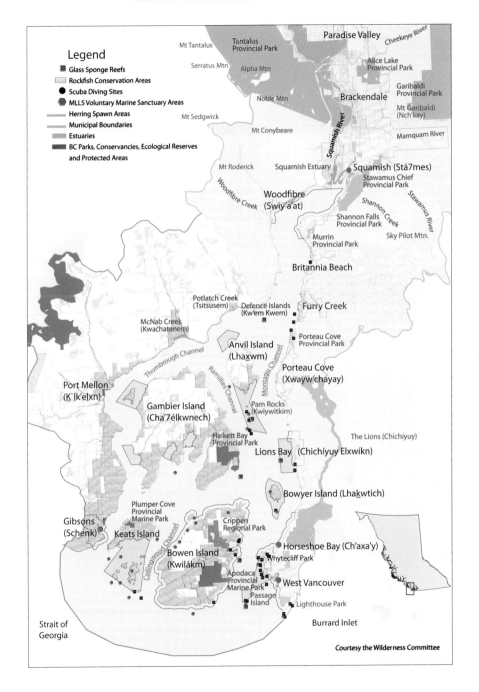

Legend

- ■ Glass Sponge Reefs
- ▨ Rockfish Conservation Areas
- ● Scuba Diving Sites
- ⬣ MLLS Voluntary Marine Sanctuary Areas
- ～ Herring Spawn Areas
- — Municipal Boundaries
- ▨ Estuaries
- ■ BC Parks, Conservancies, Ecological Reserves and Protected Areas

Mt Tantalus
Tantalus Provincial Park
Paradise Valley
Cheekeye River

Serratus Mtn
Alpha Mtn
Alice Lake Provincial Park

Noble Mtn
Brackendale
Garibaldi Provincial Park

Mt Sedgwick
Mt Garibaldi (Nch'kay)

Mt Conybeare
Squamish River
Mamquam River

Mt Roderick
Squamish Estuary
Squamish (Stá7mes)

Woodfibre Creek
Stawamus Chief Provincial Park

Woodfibre (Swiy'a'at)
Shannon Creek
Stawamus River

Shannon Falls Provincial Park

Murrin Provincial Park
Sky Pilot Mtn.

Britannia Beach

Potlatch Creek (Tsitsusem)
Defence Islands (Kw'em Kwem)
Furry Creek

McNab Creek (Kwaćhatenem)
Porteau Cove Provincial Park

Anvil Island (Lhaxwm)
Thornbrough Channel
Montagu Channel

Porteau Cove (Xwayw'chayay)

Port Mellon (K`lk'elxn)
Ramillies Channel

Gambier Island (Cha'7élkwnech)
Pam Rocks (Kwiywitkim)

Halkett Bay Provincial Park
The Lions (Chíchiyuy)

Lions Bay (Chichíyuy Elxwíkn)

Bowyer Island (Lhakwtich)

Plumper Cove Provincial Marine Park

Gibsons (Schenk)
Cripper Regional Park

Keats Island
Collingwood Channel

Bowen Island (Kwilákm)
Horseshoe Bay (Ch'axa'y)

Whytecliff Park

Apodaca Provincial Marine Park
West Vancouver

Passage Island
Lighthouse Park

Strait of Georgia
Burrard Inlet

Courtesy the Wilderness Committee

"Like finding a living herd of dinosaurs" is how one scientist described the recent discovery of 16 glass sponge reefs, thought to be extinct. Although they look like plants, the glass sponges are one of the oldest animals. Photo Diane Reid

The glass sponges are fragile, as fine as crystal, as large as a small car. They provide important habitat for fish and other ocean species, and one sponge can filter and purify up to 9,000 litres of water a day. Photo Diane Reid

Sunset over Howe Sound, a spectacular fjordal inlet in Vancouver's backyard, and traditional home of the Squamish Nation. Visitors come from all over the world to hike, sail, scuba dive, paddleboard, bike, camp, kayak, fish, ski and snowboard. Photo Rich Duncan

The great blue heron, with its primeval call, is a regular visitor to the estuaries and bays in Howe Sound, feeding on herring and other small fish along the shorelines. It is classified as a species at risk in BC. Photo Will Husby

Double-crested cormorants nest in a colony on Pam Rocks, a protected little island in Howe Sound. On land, they often adopt a distinctive posture standing upright with wings outstretched to dry their feathers after diving for fish. Photo Will Husby

Mature bald eagle guarding its nest. In the winter months, bald eagles can be viewed in the thousands along the shores of the Squamish River, feasting on salmon at the popular Brackendale Eagle Watch. Photo Rich Duncan

Female common mergansers are a regular sight at the lagoon on Bowen Island. They nest in tree holes. The mother can often be seen carrying her brood of up to four chicks on her back. Photo Will Husby

There are five species of salmon in Howe Sound. They offer food to more than 200 species, including bears, wolves, otters, seals, orcas and humans. Habitat preservation and restoration are critical if salmon are to thrive. They are among many species classified as critical. Photo John Buchanan

Orca spyhopping west of Defence Islands. After a long absence, the "killer whales" have returned to Howe Sound. They are considered an endangered species, with only 78 southern residents in the Pacific Northwest. Photo Rich Duncan

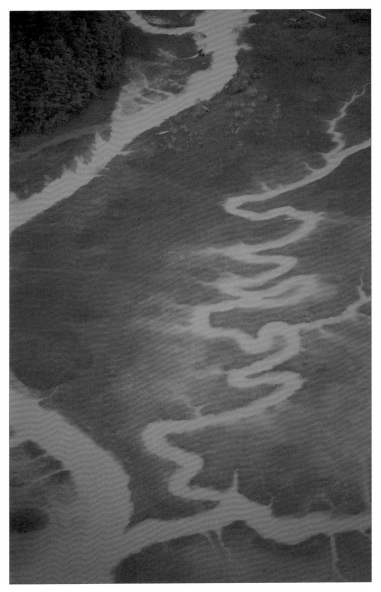

The Squamish Estuary has made life in Howe Sound possible for thousands of years. Two decades of revitalization efforts have returned large areas of the estuary to a vital wildlife habitat and reversed some of the effects of human and industrial activity. Photo Rich Duncan

Howe Sound/Atl'kitsem is a recreational paradise that is still in recovery. Less than one percent of Howe Sound is protected under provincial legislation. Photo Rich Duncan

Shannon Falls, south of the town of Squamish, is among the highest in the world, welcoming half a million visitors every year. Photo Virginia Penny

The Squamish Chief, the second-largest granite monolith in the world, is beloved by climbers from all over the world. Known to the Squamish Nation as Siyam Smanit, it has great spiritual significance and is the source of many legends. Photo Rich Duncan

The Tantalus Range with Rumble Glacier, a favourite of climbers and photographers with its impressive spires, icefalls and jagged summits. Photo Rich Duncan

109

Once devastated by logging and industrial activity, the Squamish Estuary, the jewel of Howe Sound, provides five-star accommodation for birds, bear, muskrat, deer, elk, otter and four species of salmon. Photo Rich Duncan

The Cheakamus River where the Squamish Valley and Paradise Valley meet. The name comes from Chyakmesh (people of the weir), an ancestral village of the Squamish people. Photo Rich Duncan

A small fishing boat collects prawn traps below Mount Garibaldi, BC's most famous volcano. Howe Sound has emerged as home to one of the "darlings" of the sustainable seafood movement, the spot prawn. Photo Rich Duncan

The harbour seal can hold its breath under water for up to 25 minutes. Its large eyes, protected by "oily" tears, help it to see in deep, dark waters. Photo Will Husby

In Search of
the Squamish River

"Pay attention. Attention is the only debt you owe the natural world. Pay attention.... It is the only debt that is a danger to ignore."
— Lee Maracle

It's a glorious fall day. I'm delighted to be unexpectedly travelling with citizen scientist John Buchanan to the Upper Squamish Valley. John, a preposterously fit man in his fifties, has become a bit of a Howe Sound celebrity for his videos of herring eggs and spawning salmon. He is also the first one people call when a boat sinks in the local waters.

John reminds me of Pete Seeger, the iconic folksinger who wanted to do something about the sad state of the polluted Hudson River. He restored an old sailboat and took people sailing on the river. He wanted them to pay attention to the river, to see how damaged it was and how it might be brought back to health. There is a difference: John Buchanan doesn't sing—as far as I know—and he takes videos to show the world not only the damage that has been done to Howe Sound, but also to revel in how it is recovering. Spectacularly.

The interview begins quietly enough in John's bright yellow house with splendid views of the Tantalus Range mountains. I enter the garage he's transformed into a workshop. Carefully. It's a world of sights and sounds and smells foreign to me—a jumble of pipes of assorted lengths, pieces of steel, nuts, bolts, an anvil, tools of all kinds, electrical this and that, and wire, lots of wire, all in readiness, waiting to be repurposed for one of John's

113

inventions. I perch on top of something that doesn't look too oily and get caught up in John's excitement as he talks about what he does in his workshop.

He's working on a new underwater submersible camera. The first camera he made drops down to 350 feet. He also has a pole cam, a camera that drops down 15 to 20 feet, allowing him to monitor and take photos of spawning salmon and count them, gleefully. He keeps data on the comings and goings of the spring, coho and chum, monitoring the same creeks many times in the season, and posting photos and videos on the internet. He is keenly aware of proposed developments for Howe Sound that could endanger their habitat. "I want to know the salmon story," he says. "I want to know what they make of all this."

He shows me what will soon be his pride and joy—a submersible drop camera that will be able to descend 1,000 feet into the sea. With Howey—his appropriate name for the camera—he'll be able to explore the depths of Howe Sound, to photograph the marvelous glass sponge reefs and see what kind of health rebound is happening after the Woodfibre Pulp Mill closing. He needs to be able to control the camera so it doesn't land on the bottom and harm the marine life. "I was stuck for a few days," he says. "I was missing one piece, one critical piece." The missing piece was essential for him to be able to see what he was doing. When he stumbled upon a mercury-coupling device, he was thrilled. "It cost me $150 for this little piece, but it was vital to the project."

John, the self-anointed knight of Howe Sound, rides a steed called the *Mandy Lynn*, a fifty-horsepower boat. He's a casual dresser, likes to wear T-shirts with slogans such as Save Our Earth: It's the Only Planet with Beer. He's made a name for himself as the defender of herring and salmon and all things watery. He's interested in the fact that herring are prolific in Howe Sound although they are in decline everywhere else in BC. Herring are foundational to marine food webs, providing food for seabirds, seals, dolphins, humpbacks and salmon. And their larvae feed everything from crustaceans to, surprisingly, elk. A few years ago, he snapped a photo of an elk feasting on herring roe. "As soon as you get a healthy herring population," John says, "everything benefits. The revival of herring in Howe Sound is most likely the cause of so many recent sightings of whales and dolphins."

John talks to people who don't know anything about fish, so he uses Stanley Park as a model. "I ask them if they care about the trees in Stanley Park. They do. Good. Then I tell them it's the soil that's responsible for the trees. If you don't care about the soil and it starts to be removed, eventually the trees will die. It's the same with herring. They are the soil of Howe Sound."

John has lived in Squamish since 1966. As a youth, he jigged for herring at the tip of Howe Sound, and for the last five years, he's been documenting the species, keeping track of herring spawn from Defence Island to the Squamish Harbour. "DFO (Department of Fisheries and Oceans) says herring are unpredictable. But they are so predictable. It's like clockwork," says John. "The herring find their way to shore somewhere around February the first every year, and deposit their eggs. I know where they hang out. The data are climbing—there was a spike in 2011—not necessarily because more are coming but because I'm getting better at finding them. We've also discovered resident herring."

He keeps meticulous records, noting water temperature and GPS locations, and shares his information with the DFO. His fifty-four-page report, including photos, on five years of collecting data on salmon and herring is available to anyone interested. In February of 2015, he found herring eggs along the shore very close to the proposed Woodfibre LNG site. He posted photos and videos showing abundant herring eggs on bladder kelp—commonly known as rockweed—within one hundred metres of the site.

Biologists have been warning that the proposed liquefied natural gas plant could pose a risk to herring. Project documents filed by Woodfibre in January 2015 claimed that spawning herring didn't exist within three-and-a-half kilometres of the site. Later that year a Woodfibre-commissioned herring survey verified

John's findings. It stated that the tiny fish spawn "throughout the project area," and recommended steps to prevent harm during spawning season and to enhance habitat.

John shakes his head. "The LNG plant makes no sense in this location." What also disturbs him is how the project has divided families. It reminds him of the 1970s, when a coal port was being promoted for Squamish. His father was in favour of the port facility, and his mother, who was running for mayor, wanted to find out more information before making any decision. "It was the first time I heard my parents argue." The entire community was polarized.

An anti-coal campaign was organized. The Save Howe Sound Committee was formed. Members of the Squamish Nation were also opposed. John showed me a clipping from the January 24, 1973, issue of the *Squamish Times,* quoting Chuck Billy, a Squamish Nation councillor, who appeared before the Squamish mayor and council: "Can you imagine what it would be like with coal piled here?" he was quoted as saying. "I've talked to longshoremen and you can't keep coal clean. They can't complain about it; it's their livelihood," Billy added. "The rivers and the fish are part of our livelihood now that we have destroyed the Capilano." When Mayor Brennan asked Billy if he was opposed to development, he replied: "You guys," referring to the members of council, "can leave here and head for another climate where the air is cleaner. But I plan to live here."

A delegation of the Save Howe Sound committee attended a meeting of the Vancouver City Council and expressed grave concerns about the development of a coal port in Squamish with ships passing through Howe Sound for loading. On March 6, 1973, Mayor Phillips and council unanimously passed a resolution, stating that not only should the provincial government immediately halt all plans concerning the coal port development but that the Howe Sound area should be considered primarily as a recreational area. The coal port was eventually axed for environmental reasons.

"Exactly the same thing is going on today," John says, pointing out that he's not someone opposed to industry; he has always worked in industry. "I'm not an environmentalist. I just want to make sure things are done properly. The Woodfibre LNG export facility is a direct threat to the herring stocks of Howe Sound. And let's not fool anyone. LNG is just coal wearing LNG makeup."

John works as a railway car man. He's husband to Karin and dad to a sixteen-year-old son. He's been a tradesman all his life and enjoys working with his hands. One of his many occupations includes farrier, which is how he met his wife, a horse lover. As a young man, he worked as a labourer at the Woodfibre Pulp Mill until 1982. "Six to eight feet of slime got hosed into the ocean," he says. "Herring have always been at the Woodfibre site. But with the Woodfibre Pulp Mill, the numbers plummeted. Once Woodfibre shut down, the herring returned."

He's also been a journeyman steel fabricator. Today, he's a conservationist and citizen scientist. He didn't see himself acting as the policeman of Howe Sound in his free time, "but the government dropped the ball." He's referring to the cuts to the DFO and the lack of monitoring of our precious oceans. So Buchanan took it upon himself to become the eyes and ears of Howe Sound, to pay attention to the natural world.

John has always been a passionate observer of life. When he was eleven years old, his dad had a printing business and his parents published the *Squamish Citizen*. Inspired, John published his own newspaper—the *Snooper*. When I ask him what he was snooping, he says he mostly narced on his sisters, but there were occasional items about the environment. Squamish in the 1960s was hardly a haven for environmentalism. "My parents were log-it, burn-it, pave-it people. If you talked about the environment in those days, you got your tires slashed." But he did write editorial columns about waste. He doesn't like to waste things. He prefers to repurpose them. Like the camera he is working on, which repurposes the blue pipes used for finishing concrete.

He continues to be a snooper, heading out in the *Mandy Lynn*, removing styrofoam from derelict docks, and picking up the ubiquitous plastic bags. "Garbage, garbage everywhere. If I throw a load of garbage on the road, I'll be charged, but if I throw it into the ocean, nothing happens." John also participates in the Great Canadian Shoreline Cleanup, where volunteers come out to rid the Squamish shoreline of trash.

We move to the sunny verandah to continue our chat. I'm fascinated by John's extensive knowledge of the area. When I mention I'm staying in a rustic cabin along the Cheekye River, he tells me about the serpent who lives in that river, referring to a Squamish Nation legend. "Every now and then the serpent moves and things happen. The natural world is alive."

"What about the Squamish River?" I ask. "I've looked at several maps and can't seem to find the source. And I read in one book that the Squamish is a short river."

John's face changes immediately. I've touched a nerve. He is, I assume, a normally congenial man, described by many locals as "the best" and "one of a kind," but I have clearly asked the wrong question.

"A short river? You're telling me the Squamish River is a short river? What are you doing for the rest of the day?"

"Ummm ... I have a meeting with the librarian to do some research."

"Cancel it! I'm taking you up the Squamish River so you can see for yourself that it's not a short river. And while we're there, I want to go up to the Elaho Valley and see if the Elaho Giant is still alive."

The Elaho Valley, northwest of Squamish, was the centre of controversy in the 1990s. Interfor (International Forestry Products) built a logging road in 1996 into the valley without giving public notice or engaging in public consultation. This led to a campaign to protect the old-growth forest of the Upper Elaho. By 1999, confrontations escalated in the form of road blockades, tree-sits and tree-spiking—a dangerous practice of inserting metal into the trunk of a tree to impede logging or later sawmill processing. Chief Bill Williams/Telalsemkin Siyam began a peaceful intervention, mapping out a series of Squamish Nation Wild Spirit Places/Kws Kwayk welh-aynexws.

Chief Bill, along with artist-photographer Nancy Bleck and mountaineer John Clarke, invited the public to participate in summer wilderness camping weekends in Sims Creek in the Elaho. The Uts'am Witness Program brought ten thousand people over ten years to witness the spirit and energy of the land, and the powerful medicine of this sacred place. The land became a classroom. Indigenous and non-Indigenous stood together and worked together. It was a huge awakening, which led to the decision by Interfor to stop logging the Wild Spirit Places. The forests were saved, and the places for traditional stories, for plant gathering and hunting, were now safely guarded within the Wild Spirit Places, and the area reclaimed its ancestral name, Nexw-áyantsut, meaning Place of Transformation.

The Elaho Giant has been overrun by fire numerous times. Estimated to be over one thousand years old, it is the third largest Douglas fir in the world. The hot, dry summer of 2015 resulted in major fires and John wants to see if the tree survived. "I might as well do some work," he says, "while we're out sightseeing."

I'm up for the challenge. I reschedule my library appointment for the following day and climb into John's car. We leave behind the domesticated landmarks of civilization—shops and gas stations and doughnut stores—and pass by the Chief/Siyam Smanit that stands like a guardian over the town of Squamish. "If you look closely," he says, "you can see a witch flying on her broom." I do and off we go.

I learn that the Squamish River is a glacier-fed river, home to steelhead trout and four species of Pacific salmon: chinook, coho, chum and pinks. The Squamish has four main tributaries:

the Elaho, the Ashlu, the Cheakamus and the Mamquam. There are also many creeks feeding the Squamish River, and each one offers a spawning ground for young salmonids.

As we drive, John talks non-stop about the demise of Howe Sound and how it was happening all around him when he was a child. He recalls how he used to jump the fence and break into the chemical factory on Nexen Beach. "There were no controls. Forty tonnes of mercury were discharged into Howe Sound. You couldn't take crabs out—they were poisonous. At mercury levels above .7 parts per million, things start to die. The safeguard number is .43. And this is what makes me lose sleep—the level set for the Mamquam Blind Channel was .84. Nexen was the most polluted site in Canada. Since then the land has been remediated above high tide."

To add insult to injury, effluent in the form of dioxins was discharged from the Woodfibre Pulp Mill into the Sound, creating a deadly toxic environment. "Herring have three survival techniques," John tells me. "Numbers, numbers, numbers." He believes the herring have never been entirely absent in Howe Sound, and he is participating in a study with the University of Washington to discover why the Cherry Point herring have been disappearing. "They may have moved into Howe Sound because of climate change."

John is a knowledge keeper when it comes to vital information about herring.

"Do you know how herring communicate?" he asks me.

"Do they sing?"

"No. Herring communicate by farting."

I laugh, thinking John is putting me on—until I look it up. This revealing data comes from scientists who discovered that herring create a mysterious underwater noise by farting. Researchers suspect that herring hear the bubbles as they're expelled. This helps the fish form protective shoals at night. As far as I know, it's the first-ever study to suggest fish communicate by breaking wind.

John points to the left. "See that river? That's still the Squa-mish. It's not a short river. It originates from the icefields and drains into Howe Sound." As we travel deeper into the Upper Squamish Valley, the roads become rougher, and the trees change from deciduous to coniferous. We enter a logging road and cross several narrow bridges. The rocks here are covered with thick yellow moss; old man's beard hangs from the trees. We pass by the Ashlu run-of-river project. I remember someone calling these hydro projects "ruin-of-river." John laughs. "It could be good, but we don't have people out in the field monitoring to make sure it's done properly. And most of the power is going to power plants in California."

We pass by Shovelnose Creek. "Lots of damage from log-ging in the past," he says. "It's been restored, and today there are good logging practices in place." We stop frequently so John can grab his submersible camera and check for spawning salmon. We park the car and climb down to the river, silently, so as not to disturb the fish. He drops the camera in the water, makes a few notes, and counts the coho and chum as they spawn, their bodies wriggling to the rhythm of their mating.

Hold on a minute! You mean John took you all the way to Shovelnose Creek, and he didn't tell you it's one of the most important steelhead watercourses on the coast? It just happens to be one of the best nurseries for young salmon. When they took out the big old grandmother trees years ago, it caused floods to come roaring down and wiped out some of the best spawning spots. Fortunately, some good people, people who love my fish, have been fixing things up. Did he tell you the waters come from two of my famous volcanoes? Mount Fee and Mount Cayley. Mount Cayley is known to the Skwxwú7mesh people as "Landing Place of the Thunderbird." You can see the spot where that mighty bird landed. The mountain was burned black by the Thunderbird's lightning.

Back in the car, John talks about one of his big concerns: McNab Creek, on the west side of Howe Sound, across from Anvil Island/Lháxwm. Burnco has proposed a gravel mine on the estuary. They want to dig a thirty-hectare pit, build an onsite crushing and processing plant, and produce twenty million tonnes of aggregate every year for sixteen years. McNab Creek currently hosts bald eagles, Roosevelt elk and an estuary flourishing with marine life. Burnco's consultants identified twenty-three species at risk that will be threatened by this mine, and the project is located in a very productive salmon-bearing area.

"They will remove the estuary!" John shouts. Outraged, he showed up at the Burnco meeting. "The environmentalist hired by Burnco said 'no problem no fish.' They found a little trout. I'm going out there every day to do my own study in the same study area, and I find seventy-four hundred pink salmon and eight sockeye. Any biologist worth their stuff should at least be able to repeat the findings of an amateur!"

Before 1990, he says, the government had biologists on staff and the companies would pay for them to go into the field. "After the 1983 recession, DFO began cutting costs, jobs were slashed, and companies now hire their own environmental scientists and the government pays for it. When Burnco first proposed the gravel mine, DFO said no. So Burnco returns with the same application denied by DFO." John sighs in frustration. "People are more dangerous than industry."

He points left again, a big grin on his face. "Still the Squamish River." His grin fades. "The biggest smoking gun in Squamish is the training dike built in 1971. They wanted to install seven deep-sea port facilities. Only Squamish Terminals managed to get through. The idea was to train the Squamish River to flow only on the west side. This was a shitty business plan and an environmental disaster. Coho salmon need to spend a year in creeks to gradually get used to salty water before they move into the estuary. Once they're acclimatized, they move out to the ocean. Now the coho fry have to go right out to Howe Sound and they die. In 1971, their numbers plummeted."

The ownership of the land under the Squamish Terminals has had a contentious past. The Pacific Great Eastern Railway (PGE) wanted to acquire the land to build the final leg of the railway. "PGE," says John, "bought 1,099 acres of land from the Squamish Nation in 1914. The Squamish people then had only thirty-six voting members. Many had been wiped out by smallpox. They were devastated. PGE offered them $161,419. I have a copy of the BMO cheque and the telegrams that went back and forth." He says the Department of Indian Affairs took the cheque and put the money in trust so the Squamish would only get half. "Eventually the provincial government gave the Squamish Nation replacement land for the land that was stolen from them."

I'll never forget those days when the First People were trying to understand the ways of the Newcomers. There were many who took advantage of them. They were already discouraged, watching their children, their parents, their community, so many dying from Newcomer diseases. They had no choice. They accepted whatever trinkets were offered them. In the Newcomer mind, land is property, real estate. For the First People, land was everything: identity, connection to ancestors, a pharmacy, a library of stories and songs and legends, their bible, their textbooks. Land was a gift. It could never be sold.

John is adamant about the proper care and awareness of our environment. He has observed how the lack of knowledge and lack of consideration for the processes of the natural world can lead to unintended and disastrous consequences. For example, flooding is a regular occurrence in the Squamish Valley but people want to build there. "Rivers naturally walk back and forth across the valley," he says. "One hundred years ago nobody lived in downtown Squamish. Longhouses were built on the Stawamus Reserve on the hillside. Stan Clarke Park, in downtown Squamish, used to be a lake. We called it Indian Lake."

What troubles John the most is the proposed construction of the Seventh Avenue Connector Road to allow for an easier truck route to transport pipes and lumber from the Squamish Terminals to Vancouver. "The environmental impact," he says, "would be even worse than the LNG. This bloody road is the biggest disaster happening here. They want to fill in the estuary." He says the plan makes no sense. The estuary, he says, is like a fast-food restaurant for herring, salmon, seagulls, herons and many other species. He has come up with an alternative to the Connector Road—a bridge over the Mamquam Blind Channel. He hopes the Squamish District Council will adopt it.

His vision for Howe Sound? "Start with a planning department. Hands off the estuary. This is what I want you to tell people." He looks over to make sure I'm writing all this down. "First, the biggest threat to the estuary is the training dike. It should be removed. Second, don't build a road through the middle of the estuary; build a bridge off Mamquam Blind Channel. Third, slow down development. Work with the port facility; they have to be held to the highest standards." I ask John if he thinks attitudes are changing. "We change when we're on the precipice," he says. And he's not sure we've reached that precipice yet. I only hope he's wrong.

I ask John why he's doing all this valuable scientific work as a volunteer when he could be taking it easy on his days off. His reply? "It's a lot of fun—building Howey—and I love being outdoors." And someone has to do it. He's the only one collecting data on McNab Creek. He's also monitoring salmon in Britannia Creek, the site of Britannia Mine, from which hundreds of kilograms of heavy metals flowed into Howe Sound every day for more than a century. "The acidity level on the surface of Mars," John says, "was better than in Britannia Creek." When he was a kid, the waters of Britannia Creek would burn your skin. "Today, I can't get enough of this creek. It's the black sheep in the family of North American creeks that's come back to life."

Remedial techniques have improved the quality of the waters in Britannia Creek so it's "clean enough to enjoy a swim," and John was the first one to spot salmon there in 2010. "It represents what we're capable of doing as a species." John tells me about the Millenium Plug that redirected the polluted water, adding that it was no quick and easy fix. It was given the name Millenium because it was assumed it would take a thousand years to deal with the toxins. "The facility will have to run for all time. A private developer from China wanted to build a housing development there. I went to the meeting where the biologist said there are no fish there. I stood up and said, oh yes there are. The difference is I'm not on anyone's payroll."

I look to the left at a flowing river. "Squamish River?" I ask. "Yup," answers John, laughing. Then we arrive at Mud Creek. The bridge has been washed out by recent heavy rains. There's no getting across. John is clearly disappointed. He won't be able to show me the entire length of the Squamish River. It's okay, I say. I readily admit it's not a short river. We turn back reluctantly. John is concerned about the health of the Elaho Giant. On our return, there is good news. The Giant is still standing tall. When firefighters saw it was smoking, they sprayed it continually to keep it from drying out.

"Howe Sound is a spiritual and recreational area for Vancouver, for tourists and for the seaside communities of BC. I celebrate it! All of it!" says John, his love for his home written all over his face.

Update: In May of 2016, John makes the dangerous hike to the valley, clambering over downed trees to check on the Giant, "like going out to visit an old friend to see how he's doing," he says. There he makes the sad discovery, a blackened trunk and barren branches. "The Giant was completely black. The root system extremely dry. I put my hand in it; it was all ash. There was nothing green, no green needles anywhere." He mourns the loss of the Giant. It's the loss of a beloved Elder. Here in the Squamish valley, there are still people who respect their Elders, including the venerable old trees.

WHALE IN THE DOOR

"Our future and the well-being of all our children rests with the kind of relationships we build today."
— Chief Robert Joseph, Reconciliation Canada

On March 27, 2015, the Howe Sound Science and Knowledge Workshop brought together sixty expert knowledge holders from diverse backgrounds to identify and share observations, information, data and stories related to the marine life in Howe Sound. The idea was to collect, not just statistics and numbers, but people's experience on the water, to share information and form relationships.

At the workshop, Chris Lewis, former policy advisor to the BC Assembly of First Nations and an elected councillor of the Squamish Nation, told a story of his people as told to him by his Elders: "Mink, a trickster, and his sister, Skunk, gave a big potlatch on Gambier Island. All the animals from around the area came to his longhouse. Even the undersea creatures attended, being placed in large cedar bentwood boxes filled with water. In addition to all the other species, Mink invited Whale. After everyone else was already inside, Whale swam fast and launched himself onto the land in order to peek his head into the door of the longhouse. In doing so, his large head blocked the whole entrance and trapped the others inside. This was all part of Mink's plan to get everyone together and force them to talk to each other."

"Now the whales have returned to Howe Sound after a long absence," Lewis added. "They are doing the same thing—they have trapped all of us in one room and we are talking to one another."

And we are talking to one another. We've come together because of our love for this fjord. We're talking, we're listening, we're sharing and we're forming relationships. Through information sessions, forums, actions and festivals, we are seeking new ways of being together, of living respectfully on the land and the waters, finding a way forward, a creative way, an unexpected way.

I love this story. Sometimes you need a trickster to get the plot moving. As a writer of screenplays and novels, I was trained in the storytelling device of the hero's journey. It goes like this: the hero/heroine is given a challenge, leaves on a quest, learns from mentors, vanquishes the monster/bad guys, and returns to the community as a saviour. If this scenario sounds familiar, that's because it's the template for most movies, TV shows, novels and comics. Mythologist Joseph Campbell wrote that he observed variations of this narrative structure in almost every culture he studied. The hero's journey, he believed, served as a useful story to empower adolescents and to inspire them to take on their role in the community.

I suggest it's time to move on, to grow up, to leave the hero myth behind as a quaint and perhaps useful artifact of the past. The tasks that are facing us now demand grown-up narratives—narratives that empower communities of people to come together and transform their world. Communities are not out there to slay dragons; they're there to build something new. Together. No single person can imagine what this will look like, but a community of dedicated people can dream together and design a desirable future.

The challenge in achieving a collective design for Howe Sound is both geographical and political. The Sound is a patchwork of municipalities and communities tied together by a unique geography; to travel from Bowen Island to Gibsons, for example, means taking two ferries. The jurisdictions of lands and waters within Howe Sound include First Nations; three regional districts and seven municipalities; the Islands Trust; and multiple provincial and federal government agencies, including Fisheries, Transport and the Environment. The most important jurisdiction in a moral sense is the Squamish Nation, whose traditional territory embraces the length and breadth of Howe Sound. The Musqueam Nation, the Tsleil-Waututh Nation and the Lil'Wat Nation also have traditional hunting and fishing rights in Howe Sound.

I wanted to talk to some of those people "trapped inside the room," coming together to propose a new vision for Howe Sound. I started with "the mink." Ruth Simons, a retired businesswoman, is the volunteer executive director of the Future of Howe Sound Society, a not-for-profit society engaged in the conservation and stewardship of Howe Sound. Like the mink in the Squamish legend, she lives near the waters of Howe Sound and is an excellent swimmer. Her knowledge of the history of Howe Sound is encyclopedic, and her grasp of the many issues impressive. Despite being keenly aware of the influence of corporations on governments, she continues to work optimistically to talk with all levels of government and present her case. Her sweet voice belies the steely determination of a mother bear protecting her young.

Ruth is committed to a holistic plan for the Sound. "Instead of looking at one part, one project, we need to look at the Howe Sound region as a whole. How we treat rivers and streams, how we do fishing, logging, affects everything. Everything's connected. We need a comprehensive land and marine use plan."

Over the years, Ruth has observed how "the Sound has been governed in bits and pieces. We need to have it recognized as a special place, and create a plan for all of it. We started that conversation years ago. We got the stakeholders in the room and now we're on the journey to putting the pieces together." It starts with a holistic vision of Howe Sound, Ruth says, and learning more about the region. "Building the knowledge and collective vision helps local governments plan for a more sustainable economy. For example, there are new people moving to Squamish. They are involved in recreation and technology. We want to remove barriers for people in sustainable industries."

Ruth became involved in 2008 when she was a council-lor for the municipality of Lions Bay, whose original Squamish name is Chíchiyuy El<u>x</u>wíken. Ruth learned that Burnco, an Alberta company, wanted to put a gravel mine in the McNab Creek estuary. The residents formed the Future of Howe Sound Society, and initiated a dialogue that got bigger. Today the so-ciety works with a coalition of at least twenty-three groups and its mission has been to engage stakeholders, various levels of government and First Nations through forums to discover their common values. Ruth saw how these groups could shape the future. But first "they needed to talk to each other."

The Burnco mine proposal has come up again after being turned down twice by the Department of Fisheries and Oceans. Located in the delta of ecologically sensitive McNab Valley on the shores of Howe Sound, the environmental impact is compli-cated because the mine would be right behind the estuary and would affect productive fish habitat. "We're not at peak gravel," says Ruth, "and the long term environmental and social costs of this mine far exceed any economic benefit for the region."

Ruth believes it is neccessary for all those involved to come together and make their voices heard. This is what happens at the Howe Sound Community Forums, established in 2002. Twice a year, representatives of regional districts, municipalities, feder-al and provincial officials, and the Squamish Nation engage in dialogue around the values of the Howe Sound region seek-ing common ground. Ruth finds hope that "this moving away from harmful industry to a way of restoring and giving back is demonstrated in the incredible recovery of Howe Sound."

Giving back. That's how to live well on the land. You have to understand that everything you receive from the land and the waters of Atl'kitsem is a gift. The gift of a wild salmon for supper, the gift of a cedar log, the gift of clean air. Eating a wild berry off the bush, catching a glimpse of a whale breaching in the bay, it revives you, helps you know you belong to everything. The right thing to do is to say thank you for the gift—Chen kwen mantumi—and then find a way to pay it back. That's what sustainability looks like.

~~◎~~

On April 29, 2016, the community gathered once again for another Howe Sound Community Forum. The location was Gambier Island, the largest island in Howe Sound. Eighty elected officials, members of the Squamish First Nation, government staff and NGO representatives travelled by water taxi from Pemberton, Whistler, Squamish, West Vancouver, Snug Cove, Horseshoe Bay, Lions Bay and Gibsons for a day at Camp Fircom. Gambier Island is in the traditional territory of the Squamish Nation, which includes all the islands in Howe Sound, and all the land and waters that flow down from the mountains and icefields of Whistler and Pemberton.

A warm, sunny, glorious west coast day. One that brings on instant amnesia about the amount of rain that regularly falls here. Many of us stand outside on the water taxi to take in the mountains, the islands, the beauty of Howe Sound that never ceases to amaze us, to make us feel grateful. As we step onto the dock, Gambier Island welcomes us with her beautiful beaches and forests. The island has maintained much of its wildness through dedicated efforts to protect the old forests. We climb up the hill and assemble in the light-filled meeting room, a tangible buzz of energy flowing joyfully and noisily.

Chief Ian Campbell/Xalek/Sekyu Siyam of the Squamish Nation is a hereditary chief and elected councillor. He opens the proceedings with drumming and a welcome song, the same welcome song that Chief Joe Capilano sang for King Edward Vll in 1906. I'm reminded of author Lee Maracle's words: "Our songs are prayers and these songs precede decisions. These songs remind us where our loyalties lie." I give a silent thank you to a culture that still knows the importance of song.

I've heard Chief Ian sing before, and once again I'm impressed by the beauty and power of his voice, the depth of his words.

He expresses his appreciation for these forums: "It's an exciting era of reconciliation," he says, "where we are no longer invisible on our own lands. It's exciting to be building these relationships on the shared territories of many nations. And good to hear our language spoken." This in reference to the many sincere attempts on the part of non-Squamish language speakers to wrap their mouths around the original name for Gambier Island, Cha'7elkwnech.

The Squamish Nation is at the forefront in documenting a holistic vision for Howe Sound. Over a decade ago, they completed a land use plan—Xay Temixw/Sacred Land—and are currently working toward a marine use plan. They are also involved in various development projects on their traditional territory. The territory of the Squamish Nation includes some of the most highly valued land in Canada. Their businesses include Mosquito Creek Marina, Lynnwood Creek Marina, a driving range in North Vancouver and the Capilano River RV Park. The Park Royal Shopping Centre, International Plaza, and the Greater Vancouver Storage Sewage Plant are a few examples of existing rentals on Squamish land. There are also plans to partner on the development of various parcels of land, including an apartment complex in Vancouver and the Woodfibre LNG plant on the former village of Swiy'a'at.

In 2014, the Squamish Nation signed with Woodfibre LNG to enter into a legally binding agreement, allowing Squamish Nation to conduct its own environmental assessment to determine if they should endorse the $1.6 billion proposal for a production, storage and marine transfer facility at the mouth of the Squamish River. The project will produce twenty-one million tonnes of LNG a year. Clean Energy Canada, a climate and clean energy think tank, estimates the 142,000 tonnes of carbon pollution annually will be equivalent to adding 36,000 cars to BC roads each year.

The Band Council gave a controversial environmental certificate of approval, along with twenty-five conditions, to the Wooodfibre project in August 2015. "This is one step in a multistage process, so it's definitely not a green light for the entire project," Chief Ian was quoted as saying in an interview with the *Squamish Chief.* "It allows us to issue an environmental certificate that would be legally binding. The Woodfibre LNG facility must abide by all the conditions that the Squamish Nation has imposed."

One of the main concerns was the discharge from Woodfibre's planned sea water cooling system, notably the impact that warm chlorinated water would have on the waters of Howe Sound and the all-important forage fish, salmon and eelgrass. In the fall of 2016, Woodfibre met the Squamish Nation condition and agreed to switch to an air-cooling system, which promises to be kinder to marine life.

However, this method has its own environmental drawbacks. As explained in Woodfibre's 2015 Assessment of Alternative Cooling Systems document: "The use of air coolers for the Project would require greater refrigerant inventory, increasing the potential magnitude and extent of contamination in the event of a leak or spill from the refrigerant piping or tube bundles." Other concerns with air-cooling systems include increased noise, and the need for diverting water from Woodfibre Creek in addition to Mill Creek, both fish-bearing streams. I'm learning how every solution creates its own set of problems.

On November 4, 2016, Christy Clark began her election campaign with the announcement that "Woodfibre LNG is one step closer to becoming a reality with First Nations backing and funding in place." The province chartered a boat and helicopter to transport media to the site, which has no road access. As she stood on the site of the ancient Squamish village, her trademark hard hat in place, she announced that the Woodfibre LNG was "a go," and although she claimed First Nations' support, neither Chief Ian, nor any other chiefs or members of council, were in attendance.

In an interview that same day with CBC Radio, Chief Ian responded to the announcement: "It's too early to celebrate. We're not there 100 percent. A number of issues still need to be dealt with. Squamish Nation has been very clear: we have an independent assessment. We're participating in a parallel process to the government. Nobody is going to make these decisions for us. It's going to be the Squamish Nation in response to concerns heard from our members in looking at whether or not the proponents and the Province of BC will fulfill their obligations to the Squamish Nation, which are largely centred around the environment."

Meanwhile, back at the April 2016 forum, and before concluding his remarks, Chief Ian makes a brief mention of the Woodfibre decision: "This has been very divisive in our community." His comment is greeted with a respectful silence. There is an understanding that the Squamish Nation are entitled to take control of their destiny, something denied them for many years. There is also the awareness that members of the Squamish Nation are opposed to the project.

Nine months later in an interview with Mychaylo Prystupo of the *Tyee*—an online newspaper—Chief Ian elaborated on this divisiveness, speaking about the "members who continue to demonstrate and express their legal right to protest and express their serious concerns over the potential impacts of this project.... We have a duty to ensure that our conditions, every one of them, will be adhered to. Otherwise the nation will take other remedies to ensure that our interests are being met."

He called Premier Clark's "it's a go" announcement "misleading" and "not fully respectful of the Squamish Nation process or our own authority to work with the process." Still unresolved are negotiations on a management plan to ensure wildlife protection throughout the life of the project and an impact-benefits agreement for a share of revenues that has been rumoured to exceed one-hundred million dollars' worth of jobs, land and cash.

It's easy to understand the frustration of First Nations leadership. For over a century, wealth creation has taken place at the expense of their nation. Their forests have been stripped, their fish taken, with little benefit to their members. In the words of Chief Ian, "Now part of the economic mainstream, we insist on managing our own wealth." Environmental degradation is being measured against job opportunities and possible wealth.

I look for more understanding. Taiaiake Alfred is a Mohawk philosopher and teacher. He is also director and professor of Indigenous Governance, a program he founded at the University of Victoria. In "Restitution is the Real Pathway to Justice for Indigenous People," a chapter in the book *Speaking My Truth*, he explains the differences between Indigenous and Canadian models of societal organization and governance. "Indigenous cultures and the governing structures that emerged from within them are founded on relationships and obligations of kinship relations." Their economy is based on the sustaining of those relationships, and the "perpetual reproduction of material life...on the belief that organizations should bind family units together with their land, and on a conception of political freedom that balances a person's autonomy with accountability to one's family."

He contrasts these familial and relational structures with "the liberal democratic state in which the primary relationship is among rights-bearing citizens...and in which political freedom is mediated by distant, supposedly representative structures in an inaccessible system of public accountability that has long been corrupted by the influence of corporations." To complicate matters, the British North America Act, 1867, gave the federal government authority to make laws about "Indians and lands reserved for Indians." Even though band councillors are elected by their people, they are still accountable to the Department of Indigenous and Northern Affairs.

Taiaiake doubts that these two totally different political cultures can be reconciled. "Colonial institutions and the dysfunctional subcultures they have spawned within Indigenous communities are the result of failed attempts to force Indigenous peoples into a liberal democratic mold.... One or the other must be transformed in order for a reconciliation to occur.... Any accommodation to liberal democracy is a surrender of the very essence of any kind of an Indigenous existence."

He believes the only possibility of a just relationship between Indigenous peoples and the settler society "is the conception of a nation-to-nation partnership between peoples, the kind of relationship reflected in the original treaties of peace and friendship consecrated between Indigenous peoples and the newcomers who started arriving in our territories."

The Great Peace of Montreal (La Grande Paix de Montréal) is a remarkable example of these early treaties. This was a peace treaty between New France, the Iroquois and forty other First Nations. It was signed on August 4, 1701, by the governor of New France and thirteen hundred representatives of the Aboriginal Nations. The signature page is decorated with drawings of birds and mammals, as well as signatures—a lively coming together of oral and written cultures. Implicit in this treaty was the desire to live together, to share the land, to find peaceful ways of coexisting.

In John Ralston Saul's compelling book, *A Fair Country: Telling Truths about Canada,* he describes this nation-to-nation partnership. "Over the first two hundred and fifty years of settler life in Canada, the newcomers had at best reached the level of partnership with the Aboriginals. New France, the Hudson's Bay Company, and the North West Company consciously built their place here on the indigenous ideas of mutual dependency and partnership."

The First Nations leaders weren't negotiating ownership, he writes. "They were putting on the table concepts of complex, inclusive, balanced existence on the land," following their protocols for agreements. "They wanted balanced agreements that could work for a long time, providing both parties were prepared to keep discussing and adjusting on a regular basis—usually once a year—to maintain the appropriate equilibrium. It sounds just like Canadian federalism." These agreements were not about ownership or competition or control. They were—and still are— about sustainable human relationships.

Saul makes it clear that "the French and the Hudson's Bay Company approaches had been shaped by indigenous approaches to how civilized people should deal with one another. Highly ritualized, lengthy, filled with formalized statements of purpose, these negotiations were not about clarity and completion. They were about ongoing talks and developing relationships."

The newcomers were welcomed, he says. "They were taught how to survive by the Aboriginals. How to dress. How to eat to avoid scurvy, which simply killed those who wouldn't adapt." For Saul, this partnership is the foundation of four centuries of shaping Canada. "Anyone whose family arrived before the 1760s is probably part Aboriginal.... All I am pointing out is that the constant physical intermingling between First Peoples and newcomers...had myriad effects. As a result, we have a subconscious Métis mind."

It's a persuasive argument, one that offers a powerful way for Canada to move forward by acknowledging our past. "Why," he asks, "do we continue to stumble and resist and deny when it comes to this Aboriginal role in Canada? The most obvious answer is that we don't know what to do with the least palatable part of the settler story. We wanted the land. It belonged to someone else. We took it." He believes "the persistence of the Victorian description of Canada's past lies at the heart of many of our difficulties today."

Chief Justice Murray Sinclair, who led the Truth and Reconciliation Commission, reminds us that the original people didn't oppose confederation. In the book *In This Together: Fifteen Stories of Truth and Reconciliation*, he writes that the Indigenous peoples "were ready to work together, ready to participate. Canada, in partnership with Indigenous people, was a real hope for the world...[but] Canada betrayed the loyalty, the partnership. Canada lost an opportunity to have First Nations contribute." He advises us to "go back to our original plan to work together."

Reconciliation has not been accomplished because we had a Truth and Reconciliation process. To quote the TRC Report, "Reconciliation is an ongoing process of establishing and maintaining respectful relationships. A critical part of this process involves replacing damaged trust by making apologies, providing individual and collective reparations, and following through with concrete actions that demonstrate real societal changes. Establishing respectful relations also requires the revitalization of Indigenous law and legal traditions."

Indigenous people have been shouldering the burden of reconciliation for too long. They have tried to understand our values, our laws, our ways—so different from their own values, laws and ways—and have tried to fit in. Non-Indigenous people are playing catch-up, most of us having never learned the true history of our relationship. It is up to us now to educate ourselves, to come to appreciate the wisdom of Indigenous ways, and to integrate this knowledge into our ways of thinking and being.

I accept Taiaiake's explanation of the dysfunctional governance structures that were inflicted on First Nations to try and make them fit colonial style governance. May we have the wisdom and courage to return to the promise of the original treaties of peace and friendship, and find ways to enhance our relationships without sacrificing the health of the land and the waters. Respectful collaborations and partnerships between Indigenous and non-Indigenous are the way of the future if we are ever to achieve reconciliation with each other and with the land.

People have to work on this together. If you try to go to the ocean as a single drop of water, you evaporate before you ever arrive. When you go as many drops, as a river, as a community, you are sure to arrive at the ocean. But be careful who you choose to work with. Find out if they love this place. Find out if they will defend the waters. Fiercely. Find out if they will make sure the herring have the very best home. Or do they just want to make money, take from the land, take from the waters and leave behind their mess? Work with those who live here, who are attached to the land with their hearts, not their wallets. Work with those who care about keeping the waters clean and inviting for my salmon, those who have walked my mountain trails, those who have stood at the very top and gasped in wonder at the beauty, those who have listened to the spirits of the mountains. Work with those who will protect this beauty.

Pemberton organic farmer and MLA Jordan Sturdy is a welcome figure at the Howe Sound forums. He is pleased to announce the extension of marine protection to Halkett Bay and the glass sponge reefs of Halkett Provincial Park on Gambier. This is a huge accomplishment orchestrated by many dedicated divers and organizations. Sturdy offers the gathering a cautionary note about the expected population explosion coming our way, quoting projections of another 1.1 million people moving to BC in the next twenty years, mostly to Metro Vancouver. Another compelling reason to reach agreement on how to maintain the health and well-being of Howe Sound.

Patricia Heintzman is the mayor of Squamish, a town with a population of twenty thousand. She is a potent force at these gatherings. Like many other municipalities in Howe Sound, Squamish has serious issues with derelict and abandoned boats. "There are severe gaps in our maritime laws," she tells those assembled on Gambier. Recently a barge loaded with three hundred tonnes of industrial toxins sank in the Mamquam Blind Chanel. The cleanup continues.

Many other boats have found a watery end in Squamish waters. One of the issues is the conflict of jurisdiction: on the water, the vessel is within the federal government's jurisdiction; when the boat sinks to the bottom, it's within provincial jurisdiction. This makes it difficult for the local governments who suffer the consequences.

Heintzman's municipality is also on the frontlines of development. She believes it is the responsibility of government to get talented people and entrepreneurs interested in moving to Squamish. Recently, 7Mesh Inc. began manufacturing cycling apparel. KNWN makes snowboards and skis. Since 1998, Fraserwood Timber has provided timber products and services. They were the very first timber-drying service provider on the west coast, and they employ thirty-eight people in the Squamish community.

Squamish is also emerging as a prime location for technology and innovation. Carbon Engineering has been running a pilot plant, whose "mission is to develop and commercialize technology that removes CO_2 from the air and converts it into fuel." There are also other "fueling alternatives," in the shape of two new breweries, a cidery and a winery. The cost of housing in Squamish continues to go through the roof, and they will be putting in a new homeless shelter on municipal land.

Higher education is also playing a huge role here. Quest University, Canada's first independent university, has grown over the past decade from seventy-four students in 2007 to almost eight hundred today, and has established itself as a progressive institution. As well, UBC's Clean Energy Research Centre plans to create a research facility in the city's Newport Beach oceanfront development.

Heintzman was "not pleased" with the federal government's approval of the Woodfibre LNG project proposed by Indonesian billionaire Sukanto Tonoto, who purchased the old mill site to repurpose it into a plant to liquefy and store fracked methane gas. Her council had decided it could not support the project. "Should it go through," she told the forum, "we're going to be vigilant about ensuring they are adhering to the highest of standards." Heintzman is keenly aware that her town in in a difficult position: "We have all the impact and none of the decision-making power."

"There is a great demand on tourism," she says. In 2015, more than fifty-nine thousand tourists walked through the doors of the Squamish Visitor Centre. "The Sea to Sky Highway has already reached capacity. We need high-speed rail. We need to get together with the North Shore." Her offer to host a transportation forum in Squamish is greeted with applause. "The livability of Vancouver is dependent on the natural world that is Howe Sound," she says. "The vision has to come from us."

The vision must be strong enough to deal with the current development fever. Here is a partial list of residential, commercial and tourism developments in Howe Sound announced in recent months: Westbank Sewell's Landing Development at Horseshoe Bay: 171 residential units; Britannia South at Britannia Beach: 850 to 1,000 residential units plus commercial and community amenities; Britannia Beach commercial area: doubling of the number of commercial buildings; Klahanie Resort at Shannon Falls, District of Squamish: destination resort, hotel and other amenities; Garibaldi at Squamish ski resort: all-seasons resort with 1,500 hotel rooms, 1,850 condos and 500 single-family homes; the George Hotel and luxury waterfront condominiums in Gibsons; Squamish Oceanfront Development: six phases over twenty years, including residential, commercial, hotel and many community amenities.

There are many good reasons to develop wisely and cautiously, especially along the waterfront. According to the CORI 2017 Ocean Watch: Howe Sound Report, there are many risks involved with coastal development: increased erosion, underwater noise impacting fish and marine mammals; shoreline modification disrupting natural ecosystem processes; introduced pollutants and contaminants; reduced biodiversity; reduced seafood production; degraded nearshore and intertidal habitat; increased traffic and limited access, constraining the growth of tourism and recreation.

Pamela Goldsmith-Jones is attending the forum for the first time as the newly elected Liberal MP for the Sea to Sky, although she has been a regular attender over the years. She was mayor of West Vancouver from 2005 to 2011 with a strong environmental focus. Pam is a very busy MP who divides her time between her constituents and her role as Parliamentary Secretary to the Minister of International Trade, which involves a lot of travel. She seems to have the energy and enthusiasm for all of it.

The Liberal federal government has recently given conditional environmental approval to the Woodfibre LNG project. You can feel the tension in the room, but she is greeted with courtesy and respect. "I've worked so hard to get people together," she says. "What you say and how passionate you are is helpful to me." She sees a need to focus on the health of the ocean. There has yet been no approval of the LNG carriers necessary to ship the LNG to Asia. "Marc Garneau, federal Minister of Transport, knows we had a near toxic disaster in Mamquam a few weeks ago. We need to be able to act faster on oceans."

She receives applause when she announces the reopening of the Kitsilano Coast Guard station, which had been shut down by the previous government. Although recent decisions made by her government suggest otherwise (her government's approval of Site C Dam and Kinder Morgan, which are being opposed by First Nations in BC), she insists that for the federal government, there is no more important relationship than that with Indigenous peoples. "We need to reconcile and heal."

Adam Taylor, another regular attender at the forums, is a fifth-generation Bowen Islander, a recreational scuba diver in love with the sea. He grew up on Bowen in the 1970s, the great grandson of Jacob Dorman, who homesteaded on Bowen with his family in 1895. Adam has always been fascinated with the ocean and learned to scuba dive in his early twenties. He took up underwater photography to share his observations and his passion for the many creatures he sees while diving. On his very first dive, he saw octopus in their dens in Deep Bay. To this day they remain one of his favourite species.

It's hard to love what you don't know, so once a year Adam organizes a major event on Bowen Island. He dives into the waters of Galbraith Bay and pops up with handfuls of sea urchin, sea stars and, of course, octopus to show off the beautiful ocean creatures he loves so much. I've watched young people and their parents, fascinated and somewhat afraid of the strange look and slippery feel of these colourful creatures, so unlike our land relations. I asked Adam how the sea stars and sea urchins feel about the event. "It's not their best day," he told me, but he believes it's worthwhile to allow adults and children to get up close and become acquainted with the underwater life around us.

He hasn't dropped a fishing line in Howe Sound for over a decade. "There aren't many fish left to catch. After witnessing the decline during my lifetime and comparing that to family stories going back generations, I choose to focus on local marine conservation rather than fishing." He does see more ling cod while diving, but "it will take quite some time for levels to recover to something sustainable." Adam continues to document

141

and work for the protection of the glass sponge reefs in Howe Sound. Although he is pleased with recent legislation extending the boundary of Halkett Bay Park to include the sponge reefs, he worries about the remaining Howe Sound sponges still awaiting protection. Adam is cautiously hopeful for the return of the ecosystem he loves.

Many of those I spoke to during my research were opposed to the LNG proposal, mostly for environmental reasons, but the divisiveness over this project is not limited to the Squamish Nation. It has affected every community in Howe Sound. There were many who believed the project didn't make business or economic sense, including Chris Pettingill, a software consultant with a degree in computer science. When he was chair of the Squamish Chamber of Commerce, he looked favourably upon the project. As a businessman, he appreciated the importance of providing employment to the area.

As he learned more, he decided not to back the project. "It's not well-thought-out. They need to lay it all out, be honest and discuss it. We couldn't even talk about fracking; it was off the table." The problem, he says, is that "there were separate EAO (Environmental Assessment Office) applications for the pipeline and liquefaction facility; no single EAO process included a cumulative assessment of the three related projects." The three processes being the pipeline, the liquefaction facility and the shipping. The fourth process, which was off the table, is the fracking of the gas, which would take place in Northern BC. When I ask him if he thinks LNG is clean energy, he answers, "only measuring at the smokestack. When it comes to greenhouse gas emissions, LNG is as bad as coal. When you evaluate the whole—and unavoidable—supply chain, it's often worse than coal."

As for the economic aspect, he says, "the economics didn't make sense. Studies on LNG ignore renewables. The fact is LNG can't compete long-term with renewables. The unintended consequences of the project, the damage—what industry calls the externalities—means we would get stuck paying for them."

Which is what happened with Britannia Mine when the mining companies walked away from the disaster they left behind. Taxpayers continue to pay for this cleanup.

What Chris appreciates about Squamish is the small-town atmosphere, a little bit rough around the edges. "It's self-powered, authentic," he says. "When I look out the window, it's so beautiful I have to give thanks in some way." For the future of Squamish, he would like to pay attention to the area's natural assets, to specialize, focus on a niche. He would also like to see more involvement with the Squamish Nation. "I was slow to realize the reality of First Nations. They could be involved in more partnerships. We need to leave significant areas of land available until we sort out the unceded land part. Cash as an option is not good enough to make things right."

The Woodfibre LNG project continues to be controversial; protesters from Whistler, Squamish, Lions Bay, West Vancouver, Bowen Island and the Sunshine Coast can be depended upon to show up at constituency offices and rallies to let their elected officials know they will fight this project. The opposition has increased as a result of an article in the March 4, 2017, *Globe and Mail* by Kathy Tomlinson: "British Columbia: The 'Wild West' of Fundraising."

The article reveals how "lobbyists and other power brokers are routinely buying their way into British Columbia's political inner circles by donating generously to the party in power several times a year." The *Globe* investigation found some lobbyists violating one of the few rules the province has when it comes to political donations: that donors cannot be reimbursed by clients or companies for their political contributions. "Doing so means they are giving on behalf of other entities and leaving the public in the dark about the source of the money because the contribution is listed in public records as coming from the lobbyist not the company or person that covered the cost."

A large amount of money was donated to the BC Liberals by two staff members of Woodfibre LNG—a total of $67,149 over three years—and they were reimbursed by their company. This was happening at a time when the environmental assessment was in progress. The BC Liberals are currently under investigation by Elections BC and the RCMP for accepting illegal donations from lobbyists. The trust in politics and impartial environmental assessments has reached a new low.

Marine biologist Tracey Saxby is one of the founding members of My Sea to Sky, a grassroots community organization based in Squamish opposing the Woodfibre LNG project. "Where does the natural gas come from?" she asks. "What is the cost to the communities in Canada that will suffer the consequences? What are the long-term impacts of fracking?" The scientific evidence shows that building liquefied fracked gas plants—what the industry brands as "natural" gas—will spread a destructive web of pipelines, plants, fracking sites, compressor stations and work camps across British Columbia. It seems there is little that is natural about natural gas.

On December 12, 2016, the US Environmental Protection Agency released its mammoth, thousand-page "Study of Hydraulic Fracturing and Its Potential Impact on Drinking Water Resources." The report is organized around activities in the hydraulic fracturing water cycle, which include acquiring water, mixing the water with chemical additives, injecting the fluids into the production well to create fractures, collecting the wastewater and managing the disposal of the wastewater.

The report identifies certain conditions under which impacts from hydraulic fracturing activities can be more frequent or severe, including "water withdrawals for fracturing in times or areas of low water availability, particularly in areas with limited or declining water resources...spills during the handling of hydraulic fracturing fluids and chemicals...discharge of inadequately treated hydraulic fracturing wastewater to surface water...disposal or storage of hydraulic fracturing wastewater in unlined pits resulting in contamination of groundwater sources."

In 2016, the David Suzuki Foundation, in partnership with Saint Francis Xavier University, completed the first on-the-ground measurement of methane emissions from oil and gas sites in northeastern BC, more than 80 percent of which use fracking to extract the gas. In the most thorough study ever conducted in Canada, scientists travelled more than eight thousand kilometres using vehicle-mounted gas detection instruments covering more than sixteen hundred well pads and facilities. They found that methane emissions from fracking are more potent as a climate pollutant than carbon dioxide, yet the BC government doesn't monitor emissions from fracking, and the carbon tax does not apply to methane leaks.

Recently, the government of New Brunswick extended its moratorium on fracking. Imposed in December 2014, it will be continued indefinitely, the Energy Minister announced on May 27, 2016. Also, in St John's, Newfoundland, in a report released on May 31, 2016, a panel studying fracking recommends the province continue its freeze on fracking until major questions are answered. Minister Siobhan Coady stated: "any future decisions regarding the hydraulic fracturing industry will be based on scientific evidence and *most importantly, on a social license from the Newfoundlanders and Labradorians who may be affected."* (My italics.) Quebec has also banned the process for their province. Fracking bans also exist in France, Germany and Scotland.

"We need environmental protection in place," says Tracey. "We can't fight every single project. Can't have corporations running the show." She would love to see high-speed rail from Squamish to the city of Vancouver in order to alleviate the growing pressure on the Sea to Sky Highway. She is thrilled about the idea of Squamish becoming a clean energy hub, a hub for innovation. The University of British Columbia has committed to having a campus for clean tech research on the Squamish Oceanfront lands.

"The estuaries are most under threat from development," she says. "New research has shown that estuaries are most important when it comes to carbon capture. Blue carbon is captured by living coastal and marine organisms and stored in the plants and sediment below. "Micro algae," she says, "capture more

carbon than rainforests. The estuary is like a liver. When pollutants come down to the estuary, it soaks up the pollutants and
contains them. Everything is connected. We still have so much
to learn."

Eoin Finn, also a member of My Sea to Sky, brings a host of
scientific and business credentials to his impressive and ongoing research into LNG and supertankers. He has an MBA in
business and a PhD in Physical Chemistry. He calls himself "an
accidental activist." When he first heard about the Woodfibre
LNG proposal, he decided to do some research, to look at the
consequences—economic, environmental and social—and as-
sess the hazards, risks and benefits. He has worked in the risk
assessment sector for years. He discovered that Canada has no
LNG regulations, and that international standards developed for
the siting of LNG facilities prohibit massive tankers in a location
like Howe Sound, a narrow inlet with ferry traffic, recreational
boating and nearby populations.

Eoin has made several presentations about the LNG project,
several of which I have attended. Each time I've learned more
about the complicated fracking process: breaking the bones of
the Earth so we can power our lives. "The LNG process," he
explains, "starts in the fracking fields of northeast BC, where
the gas is extracted by brute force. Chemicals, water and sand
are injected into the Earth to shatter the shale formations and
release the gas." From wells, it's piped to cleansing stations, where
CO_2 and propane are removed. The CO_2 is vented and the others
flared off. From there, the gas is sent via pipelines to a liquefaction plant, chilled to minus 162 degrees Celsius, and stored until
it is loaded onto tankers for shipment, mostly to Asia.

According to the 2017 CORI Report on Howe Sound,
almost thirteen thousand vessels entered Howe Sound in 2015,
ferries accounting for 72 percent of the traffic. The report expresses concern that the addition of large vessels from the Woodfibre
LNG or Burnco projects "can negatively affect marine life in a
variety of ways." The risks include "illegally introduced pollutants

and contaminants from bilge water, grey water or sewage release; air pollution; higher risk of fuel and oil spills; conflict between large vessels and smaller recreational vessels (there are seventeen marinas and yacht club stations in Howe Sound); dredging for ports disturbs nearshore habitat and disrupts ecosystem processes; underwater noise impacts fish and marine mammals, and risk of harmful and fatal strikes to marine mammals."

The southern resident orcas have been considered endangered—Canada's highest risk classification—since 2003. Several of its prime breeding males mysteriously died last year. "The population is very, very vulnerable," says John Ford, head of marine mammal research for the Department of Fisheries and Oceans. The whales face several threats: stocks of chinook salmon, the whales' favourite food, have been declining while industrial contaminants in the whales' bodies have been increasing, especially the levels of PCBs, which are very high. Shipping traffic is noisy, disrupting the echolocation the whales need for social contact and hunting.

The report also goes on to say that "the marine emergency response capacity is limited within Howe Sound." This is a big concern to many, and a recent spill near Bella Bella on the Great Bear Rainforest coastline was seen as a warning. On October 13, 2016, the *Nathan E. Stewart*, a tugboat pulling an empty fuel barge, ran aground near Bella Bella, spilling diesel into the water. Over one hundred thousand litres of diesel, thirty-seven hundred litres of lube oil, hydraulic oil, and spent lubricants spilled into the pristine traditional waters of the Heiltsuk people. An inept attempt to manage the spill led to the breaking of the containment booms during a storm; as a result, clam beds, the main source of income for the Heiltsuk people, were contaminated. "This 'world-class marine response' did not happen here in Bella Bella," Heitsulk Chief Councillor Marilyn Stett was quoted as saying in *Metro News*.

I heard Heiltsuk councillor Jess Housty speak at a Pull Together fundraiser on March 4, 2017. "Heiltsuk have been raising concern about tankers for decades," she told a sold-out audience.

"We can't even deal with existing traffic. It took thirty-three days to raise the tug out of the water. We can't trust the waters to provide for us anymore. It will be generations before healing is complete. It was a hit to our spirituality, our food security. Where our children play is now polluted."

The well-respected journalist Andrew Nikiforuk, recipient of numerous awards and author of *Slick Water and One Insider's Stand Against the World's Most Powerful Industry*, takes on the subject of "world-class" spill response in his article "Why We Pretend to Clean Up Oil Spills," in the *Smithsonian*, July 12, 2016. He describes the futile attempts of "blue-smocked rescuers" cleaning oil-soaked birds and the spraying of almost seven million litres of Corexit—a dispersant banned in Sweden and the UK—on BP's catastrophic spill in the Gulf of Mexico. Their work made little difference. The BP spill killed more than a million birds, and "rough estimates indicate that out of the total amount of oil spilled, BP recovered three percent through skimming, seventeen percent from siphoning at the wellhead, and five percent from burning."

The LNG carriers will have bunker fuel on board, and the tugboats needed to guide them through the inlet will most likely be powered by diesel, like the *Nathan E. Stewart*. The "illusion of a cleanup" persists to this day even though scientists have learned that "cleaning oil-soaked birds could be as harmful to their immune system as the oil accumulating in their livers and kidneys," and that Transport Canada has admitted that "it expects only ten to fifteen per cent of a marine oil spill to ever be recovered from open water." Nikiforuk addresses the "cozy relationship" between multinational oil companies who also own the corporations licensed to respond to catastrophic spills, something Robyn Allen, economist and former CEO of the Insurance Corporation of BC, described as "a flagrant conflict of interest."

"Based on the science," Nikiforuk concludes, "expecting to adequately remedy large spills with current technologies seems like wishful thinking." He suggests three things are needed for any change to happen: "Give communities most affected by a catastrophic spill the democratic rights to say no to high-risk projects, such as tankers or pipelines"; "publicly recognize there is no real techno-fix"; and for "governments and communities to properly price the risk of catastrophic spills and demand upfront multi-million dollar bonds for compensation."

Anton van Walraven of Concerned Citizens Bowen is a regular attender at the Howe Sound forums and a former organizer of volunteers with Western Canada Wilderness Committee. In July 2015 he joined a flotilla of boats in Howe Sound to honk their horns against the presence of LNG and tankers in Howe Sound. "The water is our town square," he says, "so we went to the sea in ships. We live right in the Sound and see the marine life coming back. We've all worked so hard to clean it up, and it attracts a lot of people to this region."

Many years ago, he spent a weekend at Sims Creek in the Upper Elaho as part of the Uts'am Witness Program. Anton realized that "politicians are trapped in a line of thinking, following the narrative of those who support them. I saw that we need a different story." He has devoted his days to making sure all decision-makers are in possession of the full story and of any pertinent data.

"What we have learned from this is that there is no regional planning process in place. Howe Sound is basically up for grabs. We all need to have a vote about the future of Howe Sound, in all of the communities and with all of the stakeholders. We need a planning process for the region and we need a seat at the table. We don't want these projects just being dropped on us. Every municipality in Howe Sound has passed resolutions opposing the LNG proposal."

Anton is in influential company when it comes to asking for a seat at the table. Pope Francis addressed this issue in his encyclical on the environment—*Laudato Si, Praise Be: On Caring for Our Common Home*. In a section on "Dialogue and Transparency" (paragraph 183), he writes: "Environmental Impact Assessment should not come after the drawing up of a business proposition or the proposal of a particular policy, plan or program. It should be part of the process from the beginning, and be carried out in a way which is interdisciplinary, transparent and free of all economic or political pressures....The local population should have a special place at the table."

"People are making claims that LNG is clean," Anton says, "but they haven't done the research. It takes so much energy to produce. By the time it gets to the consumer, half the energy has been used. And it's still a fossil fuel. Methane is a potent greenhouse gas. We need to create a vision for Howe Sound and the jobs will follow."

Anton believes in the importance of working for reconciliation with First Nations. "We need to understand that the way people have been living here for a long time is actually a good way to live. We have to assimilate into their way of living and acting with the Earth." He emphasizes the "yes" in Concerned Citizens Bowen's demonstrations and meetings. "We are saying yes to salmon returning to our lagoon, yes to herring, yes to dolphins finding a stable home here, yes to clean air and water."

I was curious to discover the role of fear that's generated in discussions around LNG, so I contacted Chris Erickson, who teaches political science at the University of British Columbia. Chris wrote a book called *The Poetics of Fear*, in which he traces the use of fear all the way back to Ancient Greece. In a lively conversation with Chris, I start to get a better handle on fear as a political tool.

"It's so widespread," Chris tells me, "because it works. Fear stops us from trying new stuff. Fear is used by all sides on all issues." Today, it's industry telling us "if we don't build it, no jobs, everyone suffers, while environmentalists say if we build it, the environment is degraded, people suffer, everything's covered in oil. We build, it's a disaster. We don't build, it's a disaster. They use extreme cases to make their point. It tends to limit the imagination to find another way."

I ask Chris how one can respond to fear. His response is intriguing and unexpected. "With compassion, empathy and hospitality," he answers. As an example, there's the story he just read in the newspaper about a black man and a policeman switching places for a day. "That kind of hospitality," he says, "challenges the current attitude: if you're different, you must be hostile to me. There are examples of compassion all over, but they rarely get noticed. Compassion isn't real flashy, it doesn't shoot people on the street. We get bombarded by the non-compassion stories."

I'm reminded that I have wanted to write about the compassionate exchange between West Vancouver police and Squamish Nation youth. Fear on both sides has been transformed by hospitality as they paddle together in long-distance Pulling Together Canoe Journeys, getting to know each other outside their official and often oppositional roles and assumptions. Chris believes there is always a need behind these assumptions and opinions, "and always another way of meeting that need." In response to how we meet that need, he paraphrases *The Apology of Socrates*: the only thing I know is I don't know anything.

"We don't know everything," he says. "We're always missing something important." That seems accurate to me. I question how we refer to the latest scientific theory as if it were the absolute truth about how the world works. It would be more correct to say it is the absolute best guess we have, given the research and the available data we have today. New data tomorrow will bring a new hypothesis or best guess. Chris agrees. "One can firmly believe something but we need to stay open. I probably don't know everything. Maybe there's something we have in common."

When Chris starts talking about dreamers, we find true common ground. "What the world needs now," he says, "are dreamers. We need to start thinking more creatively. There is always more than one way. Old thinking keeps us locked in the same cycle. We are stuck in the logic of fear. Stuck with the way things are. Things could be different. Easily. We need to come up with new solutions to old problems. Are we going to create new problems? Absolutely. That keeps us humble about our solutions."

In Canada, the most egregious solution we created was the answer to the so-called "Indian Problem." The solution was to exterminate Indigenous people, kill the "Indian" in the child. "It's the logic of residential schools that's the problem," says Chris. "This is different so it must be eradicated. For the churches, the option was to convert them to Christianity. The government saw them as savages; the option was to kill them. The flawed logic of the colonialist is: We know what is right. Our way is the correct way."

He calls it "the shield logic," after the terrifying shield of Achilles, which was responsible for his victories in battle. This logic has a three-stage pattern. "One: This is the way the world is. Two: You can't do anything about it. Three: It's not me, it's god, science, the market, the economy, whatever the unquestioned authority of our time. To question the statement makes you irrational."

What about the idea that LNG will save the BC economy? "Think about that assumption," answers Chris. "It comes from a sense of lack. BC isn't great and we need to be great. We're not Toronto, not the financial capital of Canada. We don't have the heritage of Montreal. What makes us think there's something wrong with this place? You have to ask who benefits from that assumption." Chris is in favour of the tourism tagline, Super, Natural British Columbia, because it contains a cautionary note. "To keep that going you can't mess with the natural stuff."

What about our role to defend, to protect the environment? Does compassion mean we just have to accept things? "We don't have to cave in," Chris replies. "We need to find creative solutions. Dreaming is dangerous. It goes against your own gut feeling about what's right. I might think it's right for me to stand up and defend, but I may be perpetuating the problem. Opposing, he says, builds in assumptions of hostility. We need to chip away at that and have compassion for all sides. The dreamer is doing something entirely different. Compassion has to be modelled. Someone has to start."

The word "compassion" is not regularly associated with meetings. Fatigue, boredom, restlessness, yes. But the Howe

Sound Community forums are imbued with compassion, as well as hope, joy and a willingness to move forward together. Besides sharing important information, the goal is to enter into viable partnerships and engagements, to come to a holistic vision for Howe Sound. It is an honour to witness this coming together around the shared love for Howe Sound.

These conversations are taking place in many other gatherings in Howe Sound. They have included film nights, flotillas, marches, aquatic forums, knowledge-holder forums, rallies supporting salmon and rallies opposing LNG. Those who gather are seeking connection, they are seeking to restore wholeness, and they are seeking a biocentric view of Howe Sound. And as they gather, they are discovering how necessary they are to each other.

Part of my workday world revolves around music. This has taught me a great deal about community. Individuals, whether notes or people, are not the main event; relationships are. The music is not made with notes. The music comes in between the notes, the relationships that are found in between the notes. Relationships between stars and galaxies, between Earth and Sun, between plants and people, between song and singer. Music also taught me that a song is not about getting to the final note, the final destination; it's about enjoying getting there.

Margaret Wheatley, who has written many books on community and leadership, believes the world doesn't change one person at a time, in spite of the ads and slogans. The world changes "as networks of relationships form among people who discover they share a common cause and vision of what's possible." Rather than worry about critical mass, she says, our work is to foster critical connections.

"Through these relationships, we will develop the new knowledge, practices, courage and commitment that lead to broad-based change." She believes "determination, energy and courage appear spontaneously when we care deeply about something. We take risks that are unimaginable in any other context."

Those of us inside Mink's Longhouse together would do well to avoid focusing on worst-case scenarios. Instead we could bring our attention to what we deeply desire for Howe Sound and continue working toward that. We might ask ourselves a few questions. What risks are we willing to take to ensure the survival and flourishing of the Howe Sound bioregion? Are we willing to open our hearts and minds, to offer hospitality, to move beyond staying the course and business as usual? Are we capable of creating and maintaining meaningful, compassionate relationships with each other? Do we have the eagerness to work toward a deep reconciliation with the land?

RE-IMAGINING ATL'KITSEM/ HOWE SOUND

"Imagine designing spring," asks Janine Benyus. The founder of the Biomimicry Institute, Janine asks herself these kinds of questions every day. "Imagine that orchestration, that timing, the coordination, all without top-down laws or policies or climate change protocols. This happens every year....In all of our creativity, have we ever done something that complex, efficient, functional, beautiful and grand?"

In her TED talk, Benyus is asking us to re-imagine how we see the natural world. She believes we have forgotten something we used to know: "that we live in a competent universe, that we are part of a brilliant planet and are surrounded by genius.... Life has been on Earth three billion years, and has learned what works and what is appropriate on the planet."

Ecosystems work to create more opportunities for life. That's what rivers do. Large dams do the opposite. "Organisms that have not figured out how to sweeten their places," says Benyus, "are no longer around to tell us about it. Life creates conditions conducive to life. It builds soil, it cleans air, it cleans water, it mixes the cocktail of gases you and I need to live." Nature has already solved many of the problems we are grappling with. "Animals, plants and microbes are the consummate engineers. After billions of years of research and development, failures are fossils, and what surrounds us is the secret to survival."

The life that abounds in Howe Sound is showing us the keys to surviving and thriving. The challenge is to find a way to meet our needs without getting in the way of the magnificent life that unfolds here every day. We humans are amazingly clever, but without intending to, we have created massive sustainability problems for future generations. The greatest chemists and biologists will never be able to imagine designing the glorious spring times of Atl'kitsem/Howe Sound. Fortunately, an ensemble of brilliant organisms is already taking care of that. But we do need to imagine how to ensure that the spring seasons we love so much will continue for generations to come.

There are so many aspects of the natural world that we are only just beginning to understand. The wise thing, I believe, is to appreciate them and protect them before they're gone. This was the purpose of a report—*Sound Investment: Measuring the Return on Howe Sound's Ecosystem Assets*—written in 2015 by Michelle Molnar. Molnar is an ecological economist and policy analyst with the David Suzuki Foundation. In conversation with Michelle, she tells me that while she received a conventional economics degree, she came to appreciate the words of Gaylord Nelson, the founder of Earth Day, who believed "the economy is a wholly owned subsidiary of the environment and not the other way around." Her work, she decided, was to build an economy that cares for the environment.

"Howe Sound is an area of regional significance," she writes, "a marine region coming back to life, but it has rarely been considered as a region. As one of the most southern sound inlets on the coast of British Columbia, it provides habitat and sheltered access to a range of species and is high in biological diversity.... Connecting to the Georgia Strait and the larger Salish Sea, the region is an ecosystem of critical importance to keeping our environment in balance."

In the conclusion to the report, she writes: "The ecosystems of Howe Sound support an incredible wealth of services. The Sound's beaches, streams, forests, wetlands and nearshore ecosystems provide residents with food, clean water, a stable climate, protection from natural disasters and a place to relax, recreate and reconnect with nature. These services underpin our health, economy and culture, yet they are not included in decision-making in any systematic manner. As these natural systems are degraded, costly investments are needed to replace the lost services of eco-systems or to rehabilitate the damaged environment."

The results of her report are compelling. It conservatively estimates the value of eleven measured services across land and marine-based ecosystems at over four billion dollars per year. She uses the word "measured" because they were only able to value a fraction of the known environmental services. "Howe Sound acts as the lungs and circulatory system for the entire Lower Mainland region, maintaining air quality, stabilizing climate and protecting communities from natural disasters. If you take down a wetland, you have to create infrastructure for flood mitigation." The highest valued services in the report were tourism, recreation and storm prevention, a particularly important service in the face of climate change.

"If you don't measure the right thing," Molnar tells me, "governments don't do the right thing." She is currently working with municipalities to show them what and how to measure, and how their natural environment saves them infrastructure costs. There are no replacement costs with natural assets. They should last in perpetuity if properly maintained.

Gibsons, a thriving seaside town in the southwest of Howe Sound, is the site of an ancient Squamish Village—Schenk.In a more recent incarnation, the town served as the location for *The Beachcombers*, a long-running sitcom about a log salvager, played by Bruno Gerussi. Molly's Reach, the restaurant where much of the drama took place, still stands and is a "must see" for tourists over forty.

Gibsons is one of the first Canadian municipalities to explore managing the natural capital in their community. The municipality wanted to find out what they needed to do in order to maintain their aquifer, which stores and filters water for free. Because they had no expertise, they approached Molnar, who worked with them in a process she calls "proof of concept," taking the ideas off the paper and putting them into practice, creating a living lab for these ideas.

"We treat it like any other engineering project, like building a road," she says. "We've completed a model for the work, restructuring how the municipal departments work together. They bring their natural assets—green spaces, aquifers, creeks, and foreshore areas—into the same asset management system as man-made infrastructure, recognizing the quantifiable values they provide to the community." The natural assets are integrated into the municipal framework for operating budgets, maintenance and regular support. Natural capital assets, they realize, are cheaper to operate, and are carbon neutral, or even carbon positive. "At the end of the day," says Molnar, "our society completely depends on the natural world, and has to respect nature's laws and limits."

This impressive report got me thinking about how Mother Earth has been working her tail off all these years to provide us with what we are now calling "natural capital assets." I don't know anyone else who works so hard for so little appreciation. Would we really rather pay for the clean air we breathe and the clean water we drink?

I met many others doing the work of re-imagining Howe Sound. Ethnobotanist Leigh Joseph is another with a great vision. I heard about Leigh from Edith Tobe, executive director of the Squamish River Watershed Society. "There's an exciting food restoration project going on in our estuary," she told me. I contact Leigh, who is living in Dawson City, Yukon, with her doctor husband and two babies. I phone her after the young ones are in bed to ask about her work to re-establish the threatened northern rice root in the Squamish Estuary.

Leigh, a member of the Squamish Nation, received her Master's degree from the University of Victoria in 2010. She studied the inter-relationship between people and plants, and the complex system of traditional foods that provided excellent nutrition for her ancestors. She foresees a day when these foods will thrive once again in Howe Sound and be part of a healthy diet for her people.

One summer Leigh worked with her aunt, Joy Joseph Mc-Cullough, for the Squamish Nation Education Department in the Squamish Valley. They created a field guide and a calendar with harvest times for traditional plants. It was here that Leigh first learned about northern rice root. "The rice root or lhásem," she tells me, "is closely related to the chocolate lily. It was cultivated in gardens up and down the coast. It's a beautiful plant. Above ground, it has a bell-shaped bloom ranging in colour from a deep purple to a chocolate brown with green highlights." Below ground are the nutritious bulbs.

The bulbs of the northern rice root were a staple food for the Squamish and other Indigenous peoples all along the Pacific Rim, says Leigh. "It provided an important source of calories in a diet rich in oil and fibre, and one of the main sources of carbohydrates." Industrial activity in and around the estuary degraded the habitat of the rice root and contaminated this and other critical food sources. There was little option but to eat the European imports—rice and potatoes. "Rice root/lhásem was easier to digest," Leigh says, "because my ancestors had eaten it for thousands of years." How does rice root taste? "It has a mild flavour," she replies, "not starchy like potato, similar to rice."

Leigh found a very small population growing in the Squamish Estuary. She hesitated to work with it, concerned that the soil might be contaminated by mercury from the chemical plant that existed in the late sixties. She tested several soil samples. "There were no traces of mercury. That put our minds at ease."

In the conclusion to her Master's thesis, Leigh writes that the iconic rice root is "an ideal candidate for restoration. In particular, each mature lhásem/rice root bulb is surrounded by many smaller bulblets, each capable of growing into a new plant." In 2009, Leigh was part of a research project in Kingcome Inlet, an isolated fjord off the coast of northeast Vancouver Island. She was working with traditional root vegetables and is happy to report that rice root is doing well in the estuary there. She sees the replanting of rice root as the first building block of many in revitalizing traditional foods, which number more than 125 native plants.

What excites Leigh about this work is "learning about the connection to place through food. My ancestors would have tended this land." More specifically, her female ancestors. "The management of plants was a woman's role," she says, "the cultivation and the drying of traditional food. The ancestors gathered all their medicines from the land." Leigh cautions me to mention that rice root is not to be harvested, because it is an endangered species in the Squamish Estuary. "Part of the success of the restoration relies on the populations of the plant being able to come back."

On June 24, 2017, I was honoured to be invited to a celebration of Squamish Nation traditional foods and medicines. Leigh, other members of the Squamish Nation, and students of Quest University organized a full day of activities in Squamish: a walk to identify and learn about local plants and how the Squamish used them as food and medicine; workshops in making salves and teas from medicinal plants; and a sumptuous and healthy feast consisting of salads of local greens, a bean dip with hemlock tips, baked salmon, moose stew, mountain goat stew, a not-too-sweet-just-right blueberry crumble, various berries made into a fruit leather, and my favourite, the very first harvest of northern rice root/lhásem in a delicious clam soup. Leigh mentioned this harvest was small because they want lhásem to have a chance to re-establish itself in its original setting.

For Leigh, this was a way of giving back to her community, sharing the knowledge she has acquired, giving her people the opportunity to eat in the healthy way of their ancestors. The meal was extraordinary, as well as the workshops, the ceremony and the singing. The celebration was part of a study funded by the Canadian Institutes of Health Research, which includes the Squamish nation of Squamish Valley and three other First Nations. A team of researchers from four universities, Elders, healers, community members and organizations are building scientific evidence to see how medicinal plants and country food can help prevent and manage type 2 diabetes. This was the highlight of the many celebrations I have had the good fortune to attend during the writing of this book. The hospitality and ingenuity of the Squamish Nation continues to inspire me.

The Squamish Nation Education Department is developing ways to continue Leigh's efforts through involvement with youth. The idea is to "reconnect Squamish youth with their land and build awareness of, and provide hands-on experience with, traditional plant foods that connect them with their ancestors, their lands and their language." Cultural Journeys, a school program at Stawamus Elementary, is a collaboration between the Squamish Nation, the Sea to Sky School District (#48) and the Parent Advisory Committee. This is "an excellent program," says Leigh, "getting kids on the land. They will dig up bulbs, recognize them, separate the bulbs and replant them, mimicking the traditional management of rice root that had been lost."

In September 2017, Leigh will commence studies for her PhD at the University of Montreal. She will be "looking at elements of healing, taking back cultural pride and control of our own health, how to build a healthy cultural practice." She sees her work as a way of "healing the land and healing our own bodies, strengthening ourselves and taking responsibility."

Leigh is making her ancestors proud. It's been one hundred years since anyone harvested lhásem in my estuary. It wasn't that long ago, there were three Squamish villages along the tidal flats. The grocery store was never far away. The Skwxwú7mesh people didn't go about digging up the ground to grow things. That would be disrespectful. They tended plants in their natural environment where the conditions were perfect for that plant to thrive. Because of this, their agriculture was invisible to the Newcomers. Like so many other good things, the Europeans were unable to see what was in front of their very eyes.

The Squamish Nation has spent thousands of years re-imagining their presence in Howe Sound, guided by their Elders, their ceremonies, their love for the land. In 2001, they created a land-use plan, Xay Temixw/Sacred Land, "to determine and describe the community's vision for the future of the forests and wilderness in their traditional territory." Through community meetings and interviews, they arrived at how the lands and resources were to be protected, managed and utilized for the benefit of present and future generations. Protection is essential. As Chief Gibby Jacob pointed out: The right to hunt and fish are not worth anything if we don't have fish, if fish can't survive in the environment."

The land-use plan describes how the Squamish Nation, over the past 150 years, "has sustained tremendous economic, political and social damage as a result of the intrusion of massive numbers of people into their territory....Both the federal and provincial governments, in contravention of their trust responsibilities, have encouraged and facilitated the illegal alienation of the lands and resources that are the subject of the Nation's aboriginal title. In so doing, both governments have...impaired the Squamish people's capacity for economic self-sufficiency while enriching the ever-increasing non-Indian society at the Nation's expense."

In Xay Temixw, the Squamish Nation asserts their "Aboriginal title to those lands and waters that constitute their traditional territory, their rights to the resources of the traditional lands and waters, and their inherent right to self-determination." It's a compelling document, the work of the Land and Resource Committee of the Squamish Nation, with the endorsement of the Chiefs and council. They were assisted by a team of consultants and experts in various fields. The project team met with as many members of the community as possible to get their views. Among the concerns voiced were the protection of sacred spaces, the management of air, water, roads, tourism, forestry and economic development.

The plan has many objectives, including the protection of archaeological sites; the securing of exclusive rights to the cultivation and harvesting of non-timber forest products; restoring the forest ecosystem structure; ensuring that the Squamish Nation receives an equitable share of economic opportunities and benefits from logging in their traditional territory; maintaining the abundance of diversity of all native species of wildlife; restoring viable species of wildlife and ensuring adequate protection of their habitat; protecting riparian areas and their functions; ensuring a clean, safe, reliable supply of water for all Squamish communities and members; the creation of jobs and economic development opportunities; and a protection strategy for animals, such as Grizzly and black bears.

"Nowhere else in the world is there major Grizzly habitat within a few hours of a major city," Chief Bill Williams is quoted as saying. "It is very important for us as a keeper of the land to make room for everybody, including the animals."

There was widespread concern about the extensive logging and tourism development that has already occurred, and the pace at which remaining wild spirit places are being lost. "The majority of the nation's traditional territory has been developed over a relatively short period of time. Only a few areas remain as wilderness. These areas are especially important as natural and cultural sanctuaries for the nation, and as places to sustain and nurture the nation's special relationship to the land."

Five areas were identified by Squamish community members as Wild Spirit Places, where clear-cut logging would be prohibited. These places will be "maintained in their natural state while allowing for a full range of traditional, cultural, spiritual and other compatible uses to provide for the continuity of the community's cultural connection to the land, while allowing for their use and enjoyment by visitors who respect and honour these areas." In Xay Temixw, the Squamish Nation reclaims their spaces as sovereign people and the right to make decisions on their territory.

Ben Parfitt, a resource policy analyst for the Canadian Centre for Policy Alternatives (CCPA), writes about the state of the forest industry in BC in his article, "The Great Log Export Drain," in the March 2017 issue of *BC Solutions*, the CCPA's newsletter. He paints a disheartening picture of forestry in BC and how the provincial government has abandoned the industry for the magical and getting-harder-to-swallow LNG pill. As he questions the wisdom of the BC government in its push for an LNG plant in Howe Sound, which "may one day employ 100 people," he finds it ironic that the LNG plant would be built on lands once occupied by the Woodfibre Pulp Mill, which closed in 2006. "Such an LNG plant would be no replacement for a forest industry that—if properly regulated—could generate thousands more high-paying jobs in rural communities."

The article goes on to reveal that since 2013, "nearly thirty-one million cubic metres of raw, unprocessed logs have been shipped out of BC. That's well over one in three logs. No previous government has sanctioned such a high level of raw exports.... Had those logs been turned into forest products here at home, BC's sadly neglected and stagnating forest industry could have employed another 4,700 people." Further, "if those exported logs had been turned into timber at BC mills, enough would have been produced to frame nearly 172,000 homes."

What is the future of forestry in Howe Sound? Is anyone re-imagining how forestry might thrive here? I contact Kim Pedersen, Technical Superintendent for Howe Sound Pulp and Paper, the one remaining mill in Howe Sound, situated in Port Mellon near Gibsons on the Sunshine Coast. HSPP began operation in 1909, producing BC's first woodfibre-based paper. Their $1.3 billion modernization and expansion in the 1990s established the company as a North American leader in mechanical paper and kraft pulp manufacture. Gone were the days of stinky, polluting pulp mills. Today they meet stringent requirements and regulations for treating effluents and keeping smokestack emissions clean.

In 2015, HSPP closed part of the operation, letting go of more than a third of their workforce, 180 workers, due to extreme drought leading to a dwindling water supply in Lake Seven, which feeds the mill's paper, pulp and power operations. The fact that people are now reading their newspapers online also contributed to the closing down of that aspect of the operation. The mill now produces specialty paper and packaging mostly for the overseas market.

Pedersen is uncertain about the future of forestry, so I contact Randall Lewis, a knowledge keeper of information on all things environmental in Howe Sound. He tells me, "Forestry was the backbone and the main economy of Squamish, and up and down the coast." Archival material at the Squamish Public Library indicates one of the region's first logging operations set up camp in the Upper Squamish Valley as early as the mid-1800s. Logging also occurred in the islands of Howe Sound at various times.

Randall's grandfather built the existing logging roads in Cheekeye. He had a Clydesdale logging operation in the Upper Squamish Valley and with the old-growth logs he harvested, he built a three-storey house for his nineteen children and a barn for over twenty Clydesdale horses. "They're powerhouses," says Randall of the horses. He explains how his grandfather could afford to pay for the house and the horses. In 1913, his grandfather and many families were removed from their reserve, Senákw, on land that is now known as Kitsilano, to make way for the railway. This land had been set aside as a reserve for the Indigenous people living there. "They gave us this land forever," says Randall. But the land was expropriated and they were put on barges and sent to Burrard Inlet and up to Howe Sound. "With the compensation he received, my grandfather bought the horses."

Today, the Squamish Nation owns Tree Forest Licence 38 (TFL 38), 219,000 hectares of land. The Clydesdales have given way to a Clark 667 log skidder for small jobs. For larger cuts, they work with other logging contractors. TFL 38 includes the watersheds of the Ashlu and Elaho rivers, and the balance of the Squamish River system north of Squamish. The majority of the land is unforested mountainous terrain and icefields.

Logging in TFL 38 began in the 1950s by various owners: Squamish Valley Timber Company, Empire Logging and Mac-Millan Bloedel. In 1995, International Forest Products (Interfor) acquired the timber tenure rights from Weldwood Canada. During Interfor's tenure, harvesting occurred at a rate of about 350 hectares per year in the form of controversial clear-cutting.

Over the years, the unregulated timber removal took its toll on the land, waters, fish and wildlife. "Mature trees are natural sponges," Randall tells me. "They act as reservoirs holding back water, releasing it slowly." Stripping the large trees from the precipitous slopes "led to slope destabilization. The land slumps and slides." With no trees to hold the water, flash floods occur, wiping out many of the fish spawning areas in the small tributaries.

When Randall speaks about the land, he continually refers to the watershed. The watershed must be seen as a whole. One cannot look at forests without considering how they are an integral part of the watershed, how they maintain the watershed.

The Forest Practices Code of BC was passed in 1995 to regulate the industry, imposing reseeding measures and restricting cut block sizes and road construction in sensitive areas, as well as protecting water supplies and damaged ecology. Dave Miller, then southern manager for Empire Logging's parent company, Weldwood, was the first to admit past logging practices were bad. "We needed a good kick in the ass," he was quoted as saying. "We raped the bloody land."

⌇

You have no idea how much sleep I have lost watching my forests ravaged and disrespected. Good thing foresters are coming to their senses. Time to turn over a new leaf. There's more than trees and board feet in the forest. You've got all kinds of birds nesting there, and insects and deer and Grizzly bears all making a very nice home for themselves. And the real action? It's happening secretly on the forest floor. Mosses and lichens and fungi. And beneath the forest carpet, enormous mats of fungi working hard to keep those trees fed and watered. Keeping them connected to each other. Talking to each other. You can't just go dragging big machines on the forest floor without disturbing a world that has taken thousands of years to create. You can't just take down a whole family of old trees and think they're not going to be missed. Sometimes the trees keep me awake at night telling me their stories.

~~~

In December 2005, Interfor sold the entire TFL 38 for $6.5 million to Northwest Squamish Forestry Limited Partnership, a company held in trust by the Squamish Nation. Sqomish Forestry LP was formed, a partnership between Garibaldi Forest Products, a privately owned forest management company, and the Squamish Nation. The sale was partly in response to the province's new Forest Revitalization Act, where 40 percent of timber was to be given back to the Crown and First Nations.

Using conventional logging, they employ many Nation members who have heavy equipment, such as skidders and excavators. They also employ non-members, and lease land at the waterfront to take wood into Howe Sound, bound for US and world markets.

Although more than half of the harvest is old growth, high on the mountain, they are also harvesting second growth. "The forest industry is not what it used to be," says Randall. "Not much old growth left. And last summer a forest fire scorched the Elaho Valley. The fire was so hot it burned right down to the rocks. It was devastating. Huge plantations of thirty-year-old trees gone. Cyprus, fir and yellow cedar." They will be replanting. The area is still very unstable and they advise people to stay away.

I ask him about the export of raw logs. "Don't like that," he says. "The hard truth is money is made faster by shipping out raw logs. We should be focusing on added value. We could create thousands of jobs overnight. We have the world markets. Let's build the infrastructure, new technology, bring in mills on wheels, laser cutting. No such thing as wood waste. Use waste to make pellets for wood stoves."

Squamish Nation also has a partnership with the Lil'Wat First Nation and the Resort Municipality of Whistler for the Cheakamus Community Forest partnership, and is moving forward on plans with the District of Squamish for a community forest, which is run like a forest, except that you have the ability to infuse community values. Squamish Nation is now the majority owner of forestry licences in their territory.

One hundred years ago, everyone living in Squamish was involved in forestry in one way or another. It was hard, physical work with good pay. The operation was mostly hand powered, with teams of horses towing the logs. A steam-powered locomotive brought the logs to the water. Historically, Squamish Nation community members were employed in all aspects of the forestry industry. But when logging jobs became scarce, they were the first to be laid off.

Some of the practices used back then caused irreparable damage to streams and rivers, but modern forestry, governed by rules and regulations, has gone a long way to making sure logging is done in a more conscientious way. Logging is still a major contributor to the local economy, providing solid, good-paying jobs for local people. And because the land has been reforested, they are now harvesting trees in areas logged sixty, seventy, even one hundred years ago. With some bold re-imagining, the forest industry could become viable again in Howe Sound.

There are some important local companies doing just that. AJ Forest Products is a specialty mill producing custom cedar timbers and poles. Triack Resources recycles wood waste into a viable energy source. They accept all kinds of wood waste, including tree branches and old lumber, as well as asphalt shingles. Some of the waste is recycled into mulch for local landscapers, while other waste is broken down into chips or pellets for use in boiler fires in nearby pulp mills.

In 2003, when the provincial government reduced the Ministry of Forests, and many of its inventory and planning functions were transferred to the new Ministry of Sustainable Resource Mangement, they took away public oversight and gave companies a lot more power to decide how logging was to proceed on public land. This enabled the companies to export large numbers of logs without processing them, resulting in unemployment and the devastation of forests. Forests are a publicly owned renewable resource. Companies should be required to process the wood in the region where they are harvested. What's needed now is some political will and action to make sure this industry remains viable.

In my conversation with Randall, I mention the insight I received two years ago when I participated in a march in Squamish. We were marching against the proposed LNG site for Howe Sound, which was being sold as a provider of jobs. As we walked along the waterfront, I saw thousands and thousands of raw logs on the beach. Jobs! Jobs! Jobs! "You're absolutely correct," says Randall. "The provincial government wants to spend eight billion dollars on Site C Dam. If they took two billion of that and put it into forestry, we could have plenty of jobs."

When Sqomish Forestry purchased the TFL from Interfor, they wisely kept Jeff Fisher on as president and forestry manager. In the December/January 2017 issue of *Logging and Sawmilling Journal*, Jeff explains the logging picture in Squamish and how it differs from such places as Prince Rupert: "Up there you wouldn't have your feller buncher [a motorized vehicle that can rapidly cut and gather several trees before felling] operating in an area, and have four hikers in spandex looking to get through the block on a trail one way, and a mountain biker looking to get through the other way." Although many of the trails were not built with any legal authority, Jeff is always looking for ways to work with the mountain bikers, including going over the maps and showing them where they are going to harvest. "If we have to harvest over a trail, we'll look at reconstructing the trail when the logging is completed, or donate money to the biking association and they can rebuild the trail."

Now that the forests are under the watch of the Squamish Nation, there has been a shift in management values to include culture and heritage, ensuring that a few old-growth trees continue to be harvested to satisfy the cultural needs of the Squamish Nation. Unlike previous logging companies who came and went, the Squamish plan to stay here forever. It's in their best interest to manage the forest in a sustainable way long term.

Randall explains that before colonization, "we managed the forests. We took down trees for our canoes and our long-houses. This helped bring in sunlight so more medicinal plants could thrive and offer more foraging for deer. We managed the watershed in a holistic way. With the clearing, we knew where the huckleberries and blueberries would be. We harmonized our society to the watershed. We honoured our sacred oath with the land. With the practice of clear-cutting, the land could no longer sustain itself."

"It takes a holistic watershed to raise a community," he says. "My Elders told me not to participate in reconciliation. Their spirit and the spirit of the land was taken from them, the land that supported us for thousands of years. Residential school, those are wounds that cannot heal. My dad wouldn't talk about it. He told me to talk about reconciliation of land, waters, rivers, watersheds. Don't carry grudges, he said. It will make you unhealthy, dark spirits in your heart. For the health of future generations, fix the land. If you want to fix things, fix the land. We need to make the spirit of the land strong again. If the land is strong, it will make our spirit strong. And future generations will receive the blessings of the land."

In *A Fair Country*, John Ralston Saul outlines the history of environmentalism, which he says was based on the Indigenous sense of belonging to place. Although Canada was "one of the first countries out of the gate on environmental questions," he writes, "people around the world are watching us with increasing suspicion. Our leaders—governmental, private-sector, even much of the academic class—continue to talk as if we were serious environmental players." But they "put more energy into their relationship with technology—a personal attachment to the idea of progress—than into their relationship with place." The Indigenous relationship is centered on place, and it takes a holistic approach.

This would suggest that the environmental movement needs to be re-imagined, that our relationship to the land must be central if we are to be successful in providing future generations with a healthy ecosystem. There is also an urgent need to re-imagine the relationship between Indigenous and non-Indigenous people in Howe Sound if we are to find ways to belong to this time and place together. Rupert Ross, a retired assistant Crown attorney for the district of Kenora, Ontario, has conducted criminal prosecutions for more than twenty remote Ojibwe and Cree First Nations and spent much of his life learning about this relationship. He has learned first-hand the challenges that we face.

In "Telling Truths and Seeking Reconciliation," a chapter in *Speaking My Truth*, Ross states: "If truly respectful relationships are to ever emerge, non-Aboriginal Canadians must come to understand that there were healthy, vibrant, and sophisticated societies on this continent at the time of contact. They must understand that it was the determined policies of assimilation, including residential schools, that were primarily responsible for the damage done to those societies and the tragedies we see today. Until that history of damage is understood, it is unlikely that the dominant society will understand why they now bear the responsibility of assisting Aboriginal people in their efforts to undo the harm that was done."

It is his view that the public perception of the cultural inferiority of Aboriginal peoples, both historically and today, must be forcefully put to rest. "While it may be understandable that European settlers, when they saw a comparative dearth of technological sophistication, assumed an absence of social and cultural sophistication as well, surely the time has come to admit how wrong that judgment was." He suggests, "the very absence of preoccupation with the technological dimension may have given traditional peoples substantially more time to dedicate their curiosity and creativity to the social, psychological and cultural dimensions instead," and that this helped them achieve a level of sophistication in these dimensions "that continues to elude the rest of us."

He believes the way to set the stage for true reconciliation is "by replacing the myth of cultural inferiority with the truth of cultural richness and diversity which, while severely damaged by every strategy of colonization, retain a sophisticated validity in today's world." By making that part of our daily consciousness, we can start building a relationship that is based on mutual respect. It seems to me that one of the greatest social and cultural achievements of Indigenous peoples is the concept that relationships are the primary context of existence. Unlike our western view of the world as a collection of independent autonomous objects, the Indigenous worldview, now supported by contemporary science, understands that relationship is everything, that we are part of one interdependent Earth community. I am not myself without everything and everyone else.

One of the most ambitious and comprehensive re-imaginings of Howe Sound is the prospect of a UNESCO biosphere designation. BC Spaces for Nature initiated this possibility a number of years ago. Now a group of motivated individuals has come together to move the possibility forward. The group is engaging with First Nations and all levels of government, as well as organizations, institutions and communities throughout Howe Sound, to explore social readiness and support for an application to the Canadian Commission for UNESCO.

This is a long-term project, demanding extensive research and collaboration. Ruth Simons of the Future of Howe Sound Society is not deterred by the length of time it will take to receive this designation. "I'm here for a long time," she tells me. "I'm not going anywhere." She recognizes that a UNESCO Biosphere Region would go a long way to fostering the dialogues necessary to strike the balance of protecting Howe Sound while allowing for sustainable development.

The World Biosphere Reserve Network currently consists of 669 reserves in 120 countries, 18 of them in Canada. The two closest to Howe Sound are Clayoquot Biosphere Reserve Region and Mount Arrowsmith Biosphere Region, both on Vancouver Island. Reserves are internationally recognized for demonstrating practical approaches to balancing biodiversity, conservation and sustainable human use of the land. Without imposing laws, rules or controls over a region, the reserves acknowledge special places as living laboratories. The goal is to promote more sustainable decision making.

Ruth Simons displays an enthusiasm for the proposal that is contagious. It's the comprehensive plan for Howe Sound she has been dreaming of, something everyone could get behind. A powerful way to say yes. Acting as a biosphere region would facilitate logistical support for ongoing attainment of goals and be led by a coordinating committee that would meet regularly. The nomination process needs provincial and federal endorsement and would be co-created with First Nations.

"A biosphere is an aspirational entity," says geologist Bob Turner, also on the committee. "It creates a way of looking at a landscape. Howe Sound would be re-imagined by putting a boundary around it. It's a philosophy. A renaming process. This would be a unique biosphere because of its proximity to a major city and because it spans the urban/wild boundary." It's a transformative dream, and like all transformations it has begun with loving and brave imaginings.

In a meeting with Member of Parliament Pamela Goldsmith-Jones on August 30, 2016, I ask what she loves about Howe Sound. Her eyes light up as she speaks about Defence Island, two small islands in the northwest area of Howe Sound. The Squamish know the islands as Kw'em Kwem. "It has a mystique," says Pam. "You can feel the past, the history. It's a sacred place. The colour of the water is different. Squamish Nation had a fish weir there; they also defended their territory from these islands." She hesitates, "You need permission to go there." Pamela tells me she was taken there by members of the Squamish Nation to talk about the importance of removing oneself from community to think. An activity she seldom has time for with her busy schedule.

"Howe Sound is what brings us all together," she says, "all the towns and islands and municipalities. We think we're so different but Howe Sound connects us all, brings us together to express our shared values about being protectors in the long term." She acknowledges the role of industry, which "has played

173

an important role in Howe Sound, bringing jobs. But today, with climate change and people's consciousness being raised, plus expected growth of Metro Vancouver, Howe Sound becomes even more special, even more important as a nature preserve."

When I ask her if Ottawa really understands what is important to the west coast, she answers, "They're getting there." And they are more concerned with the economy. "It's flat in most of Canada. Alberta is hurting, borrowing to pay off their debt. Although the majority of people in Howe Sound are opposed to the Woodfibre LNG plant, many in the rest of Canada are in favour." I understand her dilemma, and I also understand that we do not create good jobs by destroying the land and the waters. On this subject, I like to quote Jae Mather of the Carbon Free Group: "If you think the environment is not as important as the economy, you should try holding your breath while you count your money."

When I mention the UNESCO plan, Pamela goes into high gear. She has taken a lot of heat for controversial decisions made in Ottawa, including approval of the Woodfibre LNG plant in Howe Sound and the Kinder Morgan Trans Mountain Pipeline in Burnaby. "We need a plan we can all buy into," she says, "an internationally accepted sacred place." She agrees to bring the plan to cabinet. It's the vision she has been looking for.

The idea of a World Heritage Site for the Salish Sea is also being floated by the Salish Sea Trust, a loose coalition of volunteers led by Laurie Gourlay, who lives on Vancouver Island. Laurie believes this inland sea—which stretches from Desolation Sound, and includes Georgia Strait, Howe Sound, and the Strait of Juan de Fuca and Puget Sound in Washington State—is unique, "with a long history of providing food, sustenance, habitat and biodiversity for marine species and those living along its shores." He would like to see it declared a World Heritage site for its cultural, historical and natural significance. A declaration would tie together industrial harbours and fishing outposts, thirty-seven First Nations' traditional lands, 419 islands, upscale beach resorts and a network of over 3,000 marine species, 113 of them threatened. "It's about getting people to realize how special this place is," says Laurie.

Of course, the sea pays no heed to national boundaries, and without a visa or permission, it carries on its merry way south to the Strait of Juan de Fuca and Puget Sound in Washington State. Sea Legacy, a group of photographers, filmmakers and storytellers whose mission is to inspire the global community to protect our oceans, has also come on board for this initiative. "The Salish Sea is a treasure worth celebrating," says Cristina Mittermeier of Sea Legacy. "A World Heritage Site for the Salish Sea would allow us to develop a different kind of economy, to dream about ourselves, not as an industrial coastline, but a Super, Natural British Columbia." Protection for the Salish Sea would also mean protection for our fjord.

There are other proposals to protect Howe Sound being imagined and actively pursued. Underway is a feasibility study for a national park on Gambier Island, and Marine Life Sanctuaries Society is working with the Department of Fisheries and Oceans and the Vancouver Aquarium to protect the glass sponge reefs in Howe Sound. These initiatives are an indication of the powerful movement to do right by the Sound, to learn from the recent toxic history of Howe Sound, and to work to create an economy that is less destructive and more conducive to life. To ask what we would like Howe Sound to look like in thirty years. One hundred years. To ask what values will guide us, to ask if a proposed project will help or hinder that future. Protection of our Sound must come, and it will be the result of imaginative thinking, negotiation and compromise.

We are caught in between two dominant ethics. The anthropocentric sees value in the land if it can generate profit. It thrives on a disrespect of the natural world, the fundamental source of our nourishment, our security, our future. The kincentric view acknowledges all aspects of the ecosystem as family. "All my relations" includes the rocks, the trees, the soil, the water. The Howe Sound story is a story of great resilience. The challenge—and the responsibility—is to do everything we can to contribute to this resilience, to understand the intricate relationships that make everything tick, and to discover new ways of thinking and acting, based on the knowledge of our fundamental interdependence and interconnection with everything else in our world.

⚭

The most important thing is to keep those relationships alive and healthy. The relationship between eelgrass and herring, between salmon and cedar, between trees and breath. Everything has a desire to live. Your place is to act in such a way that will support life, allow it to be. Notice the leaves on the trees, how they greet you as they flutter in the wind. Notice how the birds sing their sweet daybreak song. They sing to remind you that you live in the house of Mother Earth. Act accordingly.

⚭

One of the most creative re-imaginings occurring in Howe Sound is a summer camp that enhances the relationship of young people to the Sound. Camp Suzuki is a week-long camp in August for children and young adults who want to explore the beauty of Howe Sound and develop leadership skills. The aim is to empower them to become conservation champions of Howe Sound. They focus on four themes: Squamish Nation culture and language, community organization, conservation leadership, and the Howe Sound ecosystem. Experts from the David Suzuki Foundation as well as members of the Squamish Nation and local ecologists are on hand to share their knowledge and expertise.

The camp takes place on Fircom, a large oceanfront camp on Gambier Island, dedicated to sustainability and outdoor recreation. Campers sleep in teepees or cabins, and eat a mostly vegetarian menu with many ingredients from the nearby Fircom Farm. Mornings are spent exploring the island and learning about the amazing ecosystems in Howe Sound, both from a Western-based science perspective and the traditional knowledge of the Squamish Nation. There is also time for swimming, stand-up paddle boarding, hiking and building lasting relation-ships. One of the highlights is a cultural canoe trip with members of the Squamish Nation to learn about the important Howe Sound landmarks.

Campers learn how to care for Howe Sound, and how they can bring their community together to find solutions. After camp, participants apply their newfound skills and knowledge by volunteering with a local community organization that works for the betterment of Howe Sound. All of this sounds very appealing. I am tempted to lie about my age and apply, but instead I talk to Stephen Foster, one of the visionaries for this camp.

Stephen has had a long history with camps. He attended summer camps in Ontario where he grew up, and found it was a valuable experience. He began to imagine a camp in Howe Sound. "One of the defining characteristics of Howe Sound," he says, "is that we are rich in camps. There are eight or so camps here, church based, many struggling to survive. It's an outdated model." Stephen imagined a new kind of camp, one that would engage young people, entice them "from the screen to the green." Stephen has worked as a film producer. He knows the value of story. He saw the opportunity to tell a new story about camping, to create a more current model for camps, a model that included the collective knowledge of First Nations.

"The camps I went to had First Nations names, but no one could tell me anything about the names, or what they meant." He worked hard to partner with Squamish Nation. "They've created camps of their own. They're good at it." Stephen is the Howe Sound lead for the David Suzuki Foundation, so he presented his idea to the foundation and consulted with Jeff Willis of Camp Fircom. "Jeff wrote a manifesto for the kind of camp he would like to see and what David Suzuki meant to him. Then we sold the idea to the David Suzuki Foundation."

The idea includes working with young people and the knowledge they already have, so they come to see themselves as leaders. They come to understand the challenges of leadership and learn how to be effective. "We want to create strong young voices for the ecosystems," says Stephen.

The camp started in 2015 with fifty young adults. In 2016, the first children's camp was created. It took a while but the word soon got out. This year there are eighty registered in the children's camp and a long waiting list. There will be fifteen to twenty Indigenous young people in the camp. Ideally, says Stephen, he would like to see a fifty-fifty split of Indigenous and non-Indigenous children. "Beyond all the big ideas, virtually everyone I know has zero to no sense of Indigenous people." Camp Suzuki promises to change this with the arrival of the Squamish Seagoing Canoe, the presence of the Elders, and traditional Indigenous knowledge taught through ceremony and storytelling.

Stephen has observed the transformations that occur in the camp. "For the young adults, it's the real connections they make with each other. They see that there are people out there who think like they do." The camp is challenging. Campers spend twenty-four hours in silence alone in the woods. In the debriefing after, there is lots of heavy emotion. "They really come out of themselves. You can see how hungry they are and how much they hurt. They drop their phones and really show up."

He is impressed with the motivation of the younger campers as well. "They arrive at 6:30 every morning at the waterfront, and every kid is ready to get into the water." They are still experimenting with what will work in the camp, stretching the mix of kids. "We are always evolving, discovering a different way to do summer camps."

In *A Fair Country: Telling Truths about Canada*, John Ralston Saul offers a passionate re-imagining of Canada, one that could help us to move forward. He reminds us we are not a civilization of British or French or European inspiration. "We have never been. We are a Métis nation, heavily influenced and shaped by Aboriginal ideas." Unfortunately, "the myth-makers of the late nineteenth century were busy writing out Canada's past and writing in the glory of the British Empire. The single greatest failure of the Canadian experiment, so far, has been our inability to normalize—that is, to internalize consciously—the First Nations as the senior founding pillar of our civilization."

He describes how the Indigenous roots of our Canadian civilization have influenced our taste for egalitarianism, individual rights and obligations, peacekeeping, reconciliation, inclusion, non-monolithic society, looking after one another and minority rights. Canadian courts, he writes, are far ahead of our politicians, making revolutionary judgments that actualize our history with Aboriginal people. "Our deep roots are Indigenous, and there lie the most interesting explanations for what we are and what we can be."

May the Indigenous philosophy that sees humans as part of nature—not a species chosen to master it—influence the important decisions we will be making about the future of our great teacher, Howe Sound. Atl'kitsem is teaching us how to belong, how to work together, how to build bridges to connect the communities of Indigenous and non-Indigenous peoples, how to build bridges to reconnect humans to the land. If we get this right, if we are successful in imagining and realizing a Howe Sound that continues to sustain and nourish life, long into the future, we could become a model for other communities.

Atl'kitsem is a beautiful, desirable place to live and work. There will always be another project proposal, another brilliant or inappropriate idea for how to make it better, how to make more money. There will always be conflict, community disputes, differing points of view on how to live here. What's needed is a lot of listening and understanding. There will be no quick, easy fix. We need to keep the conversation going. We have an obligation to dream, to tell our stories in order to enact a new reality, a new story.

From my hilltop window on Bowen Island/Kwilákm, I see the green finery of cedar, Douglas fir and hemlock, and behind them, the awesome Coast Mountains. I have named one of the mountains Sleeping Woman. She looks straight up into the sky.

I've been told her head is Mount Strachan, her breast has no name—yet—and her very pregnant belly is called Saint Mark's Peak. She is majestic in the winter under a blanket of snow; in the fall, serene in a veil of clouds. I spend hours watching her, photographing her, welcoming her presence in my life, the sunrise giving her a rosy glow, the sunset softening her edges, backlighting her voluptuous silhouette. The crown of her head is a ski hill, the night lights sparkling like diamonds in her hair. Is she the voice of Howe Sound who counselled, cajoled and corrected me as I put words to this story, her commanding voice compelling me to be quiet and listen?

While I was in the midst of writing and rewriting, the neighbour down the hill took out several large trees to prepare for the construction of a housing complex. With the trees gone, more of the mountaintops were revealed to me. As I scanned to the left, I saw them! Well, actually, her,. One of the Sisters—Chíchiyuy. I had been learning and writing about Chíchiyuy—formerly known as the Lions—all these months, and now one of them, the west-facing one, was in my view. And, I suppose, I was now in her view. Rebecca Duncan tells me the name for Chíchiyuy looking west is Chíchiyuy Elxwíken. While Sleeping Woman looks up dreamily at the sky, this sister looks straight down at me. I have become part of Chíchiyuy Elxwíken's landscape. May I continue to be aware of this and act accordingly, fittingly, inspired by her dignity, her respect, her ability to find peaceful solutions.

Lee Maracle, in *Memory Serves*, understands the importance of story. "Our stories remind us that we are responsible for remembering from within our original context. Remembering is a process of being fed by the past, not just my past but my ancestral past, the earth's past, and the past of other human beings. We are responsible for pulling the best threads from our past forward to re-weave our lives—together."

My desire is that this book will, in some way, be part of this re-weaving, this pulling of the threads together, that including Howe Sound's past and the ancient stories of the Squamish

Nation will encourage and enhance conversation for the future. May it be a rich, full, open-hearted, open-minded conversation, acknowledging the presence of Chíchiyuy, the magnificent mountain peaks that watch over Atl'kitsem/Howe Sound, a constant reminder that transformation is possible, that our differences can be resolved through love and care and respect.

O Siem.

AFTERWORD

Writing this book has been a somewhat unsettling experience. Stepping into unfamiliar territory usually is. My first activist march; my first ride in a Cessna 172 flying over the spectacular glaciers of Howe Sound; my first visit to Totem Hall to meet members of the Squamish Nation; my first Squamish Nation pow-wow; my first taste of traditional Squamish foods; my reading of the Indian Act and the Truth and Reconciliation Commission Report compelling me to come face to face with the devastating effects of Canada's colonialist past and present.

The book started out as an environmental story, the remarkable recovery of Howe Sound, and how this still-fragile comeback is under threat from attempts at re-industrialization. I soon realized one cannot tell that story without telling the story of the people who inhabit and are shaped by the land, and how they, in turn, shape the waters, the forests, the land. An environment is not some pristine space devoid of humans. How the people live in the environment in their own unique ways is a fascinating and necessary part of the story.

Writing about the Squamish Nation seemed, at first, presumptuous. I am not Indigenous, and although I resonate with their belief in the sacredness of the land and waters, I've been brought up in traditional non-Indigenous institutions with linear ideas of how the world works. But the story kept wanting to be told.

Then the "Voice of Howe Sound" appeared. Her voice, older than time—the voice of the mountains, the rivers, the forests, the sea—speaks truth in a way unavailable to me. When I received enthusiastic support from members of the Squamish Nation, I decided to keep writing.

Raoul Peck, director of the powerful movie *I Am Not Your Negro*, said the role of the journalist is "to be witness, interpreter, and messenger." In my attempts to take on these roles, I have often failed, blinded my own cultural stories, my biases, but the attempt has changed me. I am grateful to writer Julie Salverson, who gave me the concept of "the foolish witness," one who becomes engaged, who begins without knowing exactly how to do things, but is willing to enter into the fray, to form relationships, to fail and learn and fail again. I accept the craggy mantle of the foolish witness.

I did my best to understand Squamish Nation protocol and asked to be told when I was off the mark. There was always the risk of offending someone. I soon realized that the real risk was in missing out on the opportunity to learn something, and engage with another vibrant culture.

I was received with hospitality and generosity by the Squamish Nation. Participating in their ceremonies has helped me to understand how ceremony grounds their lives, their relationship with land and family. I see how the loss of ceremony in western culture has increased our disconnect from the Earth, as well as our anxiety. I have also become aware of, and come to appreciate, the luminous art of the Squamish Nation, the eloquence of their songs, the wisdom of their culture, the strength of their people, their humour and their connection to the land.

The words of Randall Lewis continue to inform my perspective. "Fix the land," he told me. "When the land is healthy, the people and everything else will be healthy." When the land is healed, the people will be healed. All the people. All the children. All the many astonishing creatures that have lived in Howe Sound for thousands of years. There can be no true reconciliation without reconciliation with the land. As a settler, I will settle for nothing less than to do what I can to protect Atl'kitsem, and to make reparation for the ways in which Indigenous peoples and the land have been marginalized and dispossessed over the years.

As a settler, I will settle for nothing less than to do what I can to protect Atl'kitsem, and to make reparation for the ways in which Indigenous peoples and the land have been marginalized and dispossessed over the years.

A few words on Squamish Nation stories or legends included in the book. I am acutely aware of the debate around cultural appropriation: using the art or stories of another culture without permission for one's personal gain. All of the Squamish Nation people who shared their stories with me were given the opportunity to review the piece to make sure they were accurately quoted. They responded with enthusiasm and granted me permission to include their stories in this book. Any errors are mine. I hope the reader will know my intent is to present the beauty of Howe Sound, the land and the people.

"The Sisters" (Chíchiyuy) is a wonderful story. I read it years ago in Pauline Johnson's *Legends of Vancouver*. It was told to her by Chief Joe Capilano. It's a story that has resonated for me ever since. I often look up to these lovely peaks high on the Coast Mountains to acknowledge their noble presence. Over the years, I've heard the story told by several Squamish Nation Elders and artists, including Rick Harry (Xwalacktun) at the unveiling of his magnificent sculpture, Spirit of the Mountain, at Ambleside in West Vancouver, and on the same beach, several years later, by Roy "Bucky" Baker at a canoe blessing. I heard a much longer, beautifully detailed version of the Chíchiyuy story from Rebecca Duncan, a Squamish Nation storyteller and language teacher. She gave me her blessing to include this shorter version of a most uplifting story.

The following gave me permission to include their stories:

Rebecca Duncan, Squamish Nation storyteller and language teacher.

Chris Lewis, Squamish Nation Councillor and spokesperson.

Chief Bill Williams, Squamish Nation Elder.

Gloria Nahanee, Squamish Nation Elder and traditional pow-wow dancer.

Roy "Bucky" Baker, Squamish language teacher and Squamish Nation Band Councillor.

Linda Williams, Squamish Nation Parent Advisory Council.

Joyce Williams, Squamish Nation cultural teacher.

Gwen Harry, Squamish Nation Elder.

Bob Baker/S7aplek, Squamish Nation Elder, cultural advisor and performer.

Randall W. Lewis, Squamish Nation Environmental Advisor.

Leigh Joseph, Squamish Nation ethnobotanist.

Rich Duncan

Deepening Your Connection to Howe Sound

It's been said that the best way to get people to care about a place is to take them there. The people of Howe Sound were kind enough to do just that, so that I might experience the inlet from the air, from the water, on the land. These wanderings and tours helped me to fall even more deeply in love with my home. There are many things you can do to experience for yourself the wonders of this beautiful inlet.

Go to the Mountains!

My dear friend, mountain man Robert Ballantyne, knows you can't learn about wilderness by reading about it. You have to go there. "As a hiker, I've walked and climbed the Coast Mountains for years and have experienced the powerful shapes of the dark valleys as well as the sun-struck crags and the vast glaciers that are part of the history of Howe Sound. Most people live near sea level and travel along the valley bottoms. They barely glimpse the huge alpine landscape that is all around them."

Robert offers a few words of caution: "Learn what to take with you, how not to get lost, and know when it is time to turn around and leave. Go with someone who is knowledgeable, preferably someone of wisdom, and of few words. The voice of the wilderness is easily silenced by noisy sociable people." Here are a few of Robert's suggestions to get you up to the mountains. One is even wheelchair accessible!

Yew Lake: An easy, tranquil and rewarding two-kilometre walk through subalpine meadows and old-growth forest. Drive to the end of Cypress Bowl Road. Wheelchair accessible. No dogs allowed due to the sensitive marsh ecosystem.

The Chief/Siyam Smanit: One of the most popular hikes in Howe Sound. Located in Stawamus Chief Provincial Park. Views of the Squamish Valley. Intermediate hike; three to five hours. Very busy; plan on a weekday. March to November. Trail map of Stawanus Chief and Shannon Falls Provincial Parks available at www.env. gov.bc.ca/bcparks/explore/parkpgs/stawamus/stawamus_bro-churemap.pdf

Mount Garibaldi: One of the most scenic destinations in BC, with a turquoise alpine lake and a spectacular glacier. Intermediate hike. July to October. Accessible from the Rubble Creek parking lot just south of Whistler.

CULTURAL EXPERIENCES

Audain Art Museum: Permanent collection of artwork from British Columbia. Hosts exhibits from leading museums around the world. 4350 Blackcomb Way, Whistler, 604-962-0413, audain-museum.com

Britannia Mine Museum: An underground adventure. Includes many historical buildings, over nine thousand archival photos, and three hundred maps. There are special events, changing exhibits and guided tours. You can pan for gold or climb aboard the mine train and explore the haulage tunnel. Bring a sweater. 1 Forbes Way, Britannia Beach, 1-800-896-4044, www.bcmm.ca

Kwi Awt Stelmexw: A non-profit organization formed by Squamish Peoples to strengthen all aspects of Squamish heritage, language, culture and art. They have an online map of the original Squamish names for places. OhThePlacesYouShouldKnow.com, www.kwiawtstelmexw.com

Squamish Lil'wat Cultural Centre in Whistler: The centre is a showcase for the art, history and culture of the Squamish and Lil'wat Nations. I really enjoyed my visit. You will be greeted by a welcome song, taken on a guided tour, and get the chance to make a traditional craft. I wove a cedar bracelet; it's a little wonky, but I wear it proudly. You can also try on First Nations regalia and play a hand drum. The artwork alone is worth the visit. The centre is a wonderful model of collaboration, two Nations coming together to create something of great beauty and lasting value. Perfect for kids too. 4584 Blackcomb Way, Whistler, 1-866-441-7522, email: info@slcc.ca, www.slcc.ca

Squamish Nation website: An excellent site to learn more about the culture and history of the Squamish Nation. www.squamish.net

OUTDOOR ADVENTURES

Alice Lake Provincial Park: Great family area for an afternoon of fishing or swimming or an evening stroll around the lake. Reservation recommended for overnight camping: www.env.gov.bc.ca/bcparks/reserve/

BC Ferries: Take a ferry to Gibsons or Bowen Island. Even a short ferry ride is breathtaking — as scenic as any Alaska Cruise, just shorter. For schedules, go to: www.bcferries.com/schedules

Bowen Island Kayaking: Right by the ferry dock: offers tours and rentals. Bowen Island's calm waters, isolated coves and greenery make it the perfect start for an exploration of the islands of Howe Sound. 604-947-9266, www.bowenislandkayaking.com

Bowen Island Tours: Guided nature tours, historical tours and food tours. 604-812-5041, info@bowenislandtours.com, www.bowenislandtours.com

Brackendale Eagle Watch: From mid-November to mid-February, bald eagles can be viewed in the thousands along the shores of the Squamish River between Squamish and Brackendale. Attracted by spawning salmon, they congregate in the trees and feast along the shore. It's an awesome sight. Bring the kids and the binoculars. Visit or phone Squamish Adventure Centre, 604-815-4994, toll free 877-815-5084, or visit their website: www.explor-esquamish.com

Crippen Regional Park, Bowen Island: A short ferry ride from Horseshoe Bay. Start at the Visitor Centre and get a map.

Oceanfront beaches and forest trails. Meander through Davies Orchard, the Crippen Park meadow, stop by the Fish Ladder, admire Bridal Veil Falls, walk around Killarney Lake or climb Mount Gardner. www.metrovancouver.org/services/parks/parks-greenways-reserves/Crippen-regional-park

Gibsons Public Market: This is an indoor/outdoor adventure, a new gathering place in the heart of Gibsons. Rebuilt from the old yacht club building, and adjacent to the Gibsons Marina, it is a miniature version of the Granville Island Public Market. It's a food adventure, where you will find a bakery, a café, a delicatessen, fresh market veggies, cheeses and flowers. There is also an interactive Marine Education Centre where adults and young people can explore the marine life around them. www. gibsonspublicmarket.com

Howe Sound Geo Tour: "Explore the geology and landscape along highway 99 with your own geologist-in-your-pocket." You can download a pdf and take the tour created by Bob Turner and his colleagues. www.publications.gc.ca/collections/collection_2010/nrcan/M4-83-7-2010-eng.pdf

Kayak the Marine Park Trail in Howe Sound: To get you in the mood, watch the video about a couple of old farts who express their love for the Sound as they kayak the fabulous Marine Park Trail. www.youtube.com/watch?v=olliN_IAwA4

Kwemkwemshenam Cultural Tours: The North Trail Estuary cultural tour, an easy walking tour, includes traditional Skwxwú7mesh songs. The tour takes you through the former village of Skwelwil'em, now known as the Squamish Estuary Wildlife Management Area. Duration 2–3 hours. Tour season: August to October, Friday through Sunday. Contact Chrystal Nahanee 1-778-628-8165, www.firstnationstours.wordpress.com

Sea To Sky Air: Get really high with a flight over Howe Sound. Many possible flight experiences, including flying the plane yourself. These people are passionate about aviation and Howe Sound. Ruth Simons, my companion on the flight, described it as "an amusement park ride for adults." I agree. It's also a wonderful way to be in the mysterious presence of the glaciers, mountains, valleys and waters of Howe Sound. Sea to Sky Air received the 2016 Small Business BC, People's Choice Award. At the Squamish airport about ten minutes from downtown Squamish, 46041 Government Road, Squamish, www.seatoskyair.ca, 604-898-1975

Sea to Sky Gondola: Get a bird's-eye view of the Sound. A ten-minute ride in a gondola with floor-to-ceiling glass windows takes you 2,900 feet above sea level. For those who are squeamish about heights, face the mountain and don't look down. There is also a soft red blanket to throw over your head if you really get the heebie-jeebies. When you get to the top, you can choose from easy to advanced hikes. The Panorama Trail is a wide, easy, pleasant walk through open forest with spectacular views. You can also walk across the Sky Pilot Suspension Bridge for a panoramic view, or have lunch or a drink at the Summit Lodge. Located an hour from downtown Vancouver and just south of Squamish. Highly recommended. 604-892-2551, www.seatoskygondola.com

Sewell's Marina: Get on a boat and experience Howe Sound from the water. Boat rentals, sea tours, fishing charters from Horseshoe Bay/Ch'axa'y 6409 Bay Street, West Vancouver, 604-921-3474, www.sewellsmarina.com

Shannon Falls Provincial Park: just off the Sea to Sky highway, south of Squamish, a popular day-use park for hiking and picnicking. A concession, picnic grounds, an information centre, and close-up views of the falls.

Squamish Adventure Centre: the best place for information about hikes in the Squamish area. Local guide books and maps. 38551 Loggers Lane, 604-815-5084

Squamish Oceanfront Trail: an easy half-hour walk where you can view Shannon Falls, as well as kite boarders and paddle boarders in the Squamish Harbour. At the end of Cleveland Avenue, turn left on Vancouver Street, then right on Galbraith until the signed trailhead.

Sunwolf Rafting: Summer whitewater rafting, from family adventures to extreme whitewater expeditions. Winter eagle viewing floats. 70002 Squamish Valley Road, Squamish, 604-898-1537, www.sunwolf.net

Whale sightings: If you see a whale, report it by phone at 1-866-I-SAW-ONE, email: sightings@vanaqua.org or online at www.wildwhales.org

Zoom Zoom Bowen: Rent a scooter from this carbon-positive scooter rental company and explore Bowen Island. 604-725-9207, www.zoomzoombowen.com

GROUPS ACTING TO PROTECT HOWE SOUND

Bowen Island Conservancy: www.bowenislandconservancy.org

BC Spaces for Nature: www.spacesfornature.org

Bowen Island Fish and Wildlife Club: www.bowenhatchery.org

Camp Suzuki: Camp Fircom on Gambier Island hosts camps for children and young adults to engage with members of the Squamish Nation, learn about environmental leadership, and learn how to protect Howe Sound. Experiential and life-changing. Swimming, paddling, storytelling, outdoor classroom, learning on the land. www.campsuzuki.org

Canadian Parks and Wilderness Society: A national campaign to create new Marine Protected Areas. www.cpawsbc.org

Concerned Citizens Bowen: A not-for-profit society protecting the waters of Atl'kitsem/Texwnéwets'/Howe Sound. Great photos of returning whales and a video of the exquisite glass sponge reefs. www.ccbowen.ca

CORI, Vancouver Aquarium: www.vanaqua.org/act/research/coastal-ocean-research-institute

David Suzuki Foundation: www.davidsuzuki.org

Gibsons Alliance of Business and Community Society: www.gibsonsalliance.ca

Islands Trust: Working to protect communities, culture and the environment in the Gulf Islands. The Islands Trust has recently begun reconciling with and collaborating with First Nations who have historical connections and interest in the islands. www.islandstrust.bc.ca

Marine Life Sanctuaries Society: "Making the world a better place, one no-take marine protected sanctuary at a time." www. mlssbc.com

My Sea to Sky: A grassroots volunteer organization concerned about the impacts of Woodfibre LNG in Howe Sound. Lots of information and videos. www.myseatosky.org

Outdoor Recreation Council: Fostering the responsible use of BC's outdoors, and building bridges between outdoor recreation groups. Founded in 1976, representing more than 100,000 individuals, celebrating BC Rivers Day and compiling the Endangered Rivers List for BC. www.orcbc.ca

Pacific Salmon Foundation: Founded in 1987, a federally incorporated non-profit charitable organization dedicated to the conservation of wild Pacific salmon and their natural habitats in BC and the Yukon. www.psf.ca

Salish Sea Marine Sanctuary and Coastal Trail: www.salishsea. org

Salish Sea Trust: www.salishseatrust.ca

Sea Legacy: Website includes a beautiful and informative video about the Salish Sea. www.wearethesalishsea.eco

Sea to Sky Outdoor School for Sustainable Education, Gibsons: www.seatosky.bc.ca

Sierra Club BC: Inspiring generations to defend nature and confront climate change so families, communities and the natural world can prosper together. www.sierraclub.bc.ca

Squamish Nation: www.squamish.net

Sustainable Howe Sound: Changing perceptions through telling the stories. Videos about the people and the natural world of Howe Sound: www.sustainablehowesound.ca

Squamish River Watershed Society: www.squamishwatershed. com

Squamish Streamkeepers Society: www.squamishstreamkeepers.net

Sunshine Coast Conservation Association: www.thescca.ca

The Future of Howe Sound Society: A not-for-profit society devoted to preserving the natural beauty and important ecosystems in Howe Sound. Important up-to-date information: www. futureofhowesound.org

The Pacific Streamkeepers Federation: www.pskf.ca

Voters Taking Action on Climate Change: vtacc.org

West Vancouver Streamkeepers: www.westvancouverstream-keepers.ca

Wilderness Committee: www.wildernesscommittee.org

Rich Duncan

RECOMMENDED VIEWING

Biomimicry: Janine Benyus, Ted Talks, two informative and lively talks about nature's engineers: www.ted.com/speakers/janine_benyus

Eighth Fire: A powerful CBC series hosted by Wab Kinew and a team of Aboriginal storytellers from across Canada. The title is drawn from an Anishinaabe prophecy that declares now is the time for Aboriginal peoples and the settler community to come together and build the "8th Fire" of Justice and harmony. Available in libraries or for purchase from CBC.

First People of the Pacific Northwest: DVD producer Helmut Manzl, Squamish Historical Society 11/09/09. Copy at the Squamish Library.

Give the Love to the Wiggly Fish: www.youtube.com/watch?v=OydljDaLmtw

How to say Skwxwú7mesh (Squamish): www.youtube.com/watch?v=yknmoz9PZRU

Howe Sound Crest Trail: www.youtube.com/watch?v=rpTCFsqrXNg

Howe Sound: Vancouver's Wild Neighbour: www.youtube.com/watch?v=olliN_IAwA4

John Borrows: Indigenous Law 340: An informative video series of his lectures at the University of Victoria. I considered the lectures a necessary education. You could start with this one: Lecture 2: Governance: Canada's Indian Act: www.youtube.com/watch?v=3lgrZCBIwwA

Let's Talk About LNG: www.youtube.com/watch?v=AV2c5ax-BOcg

LNG in a Nutshell: A two-minute glimpse of the inefficient process that is LNG and the extractive economy, in general: www.youtube.com/watch?v=35BUwQ_BHsI

McNab Creek: A video by Bob Turner: www.youtube.com/watch?v=cMyLcJ_CSvw#t=31

Orcas on the Hunt! Howe Sound: Another great YouTube video by geologist/filmmaker Bob Turner. Two pods of transient orcas hunting a sea lion, with Bob's brother Tim in the centre of the action: www.youtube.com/watch?v=3NCuLawvQaE

Practical Decolonization: Taiaiake Alfred, Mohawk scholar: www.youtube.com/watch?v=pq87xqSMrDw

Reclaiming the Honorable Harvest: Robin Kimmerer, Ted Talk: www.youtube.com/watch?v=Lz1vgfZ3etE

Spirit of the Coast Salish: DVD, producer Helmut Manzl, Squamish Historical Society. I found a copy at the Squamish Library.

The Great Howe Sound Recovery: Five short films about the beauty and the fragility of Howe Sound. A must-watch if you want to know more about Howe Sound: www.sustainablehowesound.ca

The Revolution Has Begun: Christi Belcourt, Keynote at Laurentian University, 2016: www.youtube.com/watch?v=XqBXD-PzyLm0

The Salmon Are Back! Bowen Island BC: A lively and informative video by Bob Turner about the amazing return of Chum Salmon to Bowen Island in the fall of 2016: www.youtube.com/watch?v=15sCGelrlKU

Voices in the Sound: A musical about the history of Bowen Island, written and produced by Pauline Le Bel and performed by island actors, singers and musicians: www.youtube.com/watch?v=xOHITv3UGfQ&t=98s

BIBLIOGRAPHY

(This list will never be complete; I'm still reading and learning.)

A Fair Country: Telling Truths about Canada, John Ralston Saul, 2008, Viking Canada, Toronto, ON. "We are not a civilization of British or French or European inspiration. We have never been. We are a Métis Nation, heavily influenced and shaped by Aboriginal ideas. That we strangely fail to recognize this holds our country back."

A River Never Sleeps, Roderick Haig-Brown, 2014, 1974, Skyhorse Publishing, New York, NY.

A Voyage of Discovery to the North Pacific and Round the World 1791-1795, Captain George Vancouver, 1798, London, England.

Ancient Pathways, Ancestral Knowledge: Ethnobotany and Ecological Wisdom of Indigenous Peoples of Northwestern North America, Volume One and Volume Two, 2014, Nancy J. Turner, McGill-Queen's University Press, Montreal, PQ.

Around the Sound: A History of Howe Sound — Whistler, Doreen Armitage, 1997, Harbour Publishing, Madeira Park, BC.

Becoming Intimate with the Earth, Pauline Le Bel, 2013, Collins Foundation Press, Santa Margarita, CA.

Blackfoot Physics: A Journey into the Native American Worldview, David Peat, 4th Estate, 1996, London, UK.

Bright Seas, Pioneer Spirits: A History of the Sunshine Coast, 2009 Betty Keller and Rosella Leslie, TouchWood Editions, Victoria, BC.

Capilano: The Story of a River, James W. Morton, 1970, McClelland and Stewart Ltd., Toronto, ON.

Coast Salish Essays, Wayne Suttles, 1987, Talonbooks, Vancouver, BC.

Conversations with Khatsahlano, 1932-1954, Major J.S. Matthews, Compiled by the City Archivist, Vancouver, BC.

Diving Howe Sound Reefs and Islands, Glen Dennison, 2012, Glen Dennison.

Edge of the Sound: Memoirs of a West Coast Log Salvager, Jo Hammond, Caitlin Press, Halfmoon Bay, BC.

"EPA's Study of Hydraulic Fracturing for Oil and Gas and Its Potential Impact on Drinking Water Resources," United States Environmental Protection Agency: www.epa.gov/hfstudy

First Dollars: Pipelines, Ports, Prisons and Private Property, Bold Aboriginal Entrepreneurs are Reinventing Canada's Economy, Alex Rose, ebook, 2014, available on Kindle and Kobo.

First Fish, First People: Salmon Tales of the Pacific Northwest, Ed. Judith Roche and Meg McHutchison, 1998, UBC Press, Vancouver, BC.

First Nations 101: Tons of Stuff You Need to Know about First Nations People, Lynda Gray, 2016, Adaawx Publishing, Vancouver, BC.

Honouring the Truth, Reconciling for the Future: Summary of the Final Report of the Truth and Reconciliation Commission Canada. Download at: www.trc.ca/websites/trcinstitution/File/2015/Honouring_the_Truth_Reconciling_for_the_Future_July_23_2015.pdf

In This Together: Fifteen Stories of Truth and Reconciliation, Ed. Danielle Metcalfe-Chenail, Brindle and Glass Publishing, Victoria, BC.

Khot-La-Cha: The Autobiography of Chief Simon Baker, compiled end edited by Verna Kirkness, 1994, Douglas and McIntyre, Vancouver, BC.

Legends of Vancouver, E. Pauline Johnson, 1911, new edition, David Spencer Limited, Vancouver, BC.

Living on the Land: Indigenous Women's Understanding of Place, edited by Nathalie Kermoal and Isabel Altamirano-Jiménez, 2016, AU Press, Edmonton, AB.

Love of Country: A Journey Through the Hebrides, Madeleine Bunting, 2016, Granta Books, UK.

Memory Serves: Oratories, Lee Maracle, 2015, NeWest Press, Edmonton, AB.

The Native Voice, Eric Jamieson, 2016, Caitlin Press, Halfmoon Bay, BC.

"Notes on the Cosmogony and History of the Squamish Indians of British Columbia." Charles Hill-Tout, published 1897, www.archive.org/details/cihm_15221

Ocean Watch: Howe Sound Edition 2017, edited by Karin Bodtker. All the most up-to-date information about the health of the marine areas in Howe Sound, with many photos, diagrams and Squamish Nation history. Pdf of the report available for free download through Vancouver Aquarium, Coastal Ocean Research Institute (CORI): oceanwatch.ca

Peace, Power, Righteousness: An Indigenous Manifesto, Taiaiake Alfred, 1999, Oxford University Press, Canada.

Picturing Transformation: Nexw-Áyantsut, Nancy Bleck, Katherine Dodds and Chief Bill Williams, Figure 1 Publishing, 2013, Vancouver, BC. A beautiful picture book about the Uts'am Witness Program, a peaceful logging intervention by Squamish Nation and non-Indigenous allies.

Plant Technology of First Peoples in British Columbia, Nancy J. Turner, 1996, Royal BC Museum, Victoria, BC.

"Restitution is the Real Pathway to Justice for Indigenous Peoples," Taiaiake Alfred, a chapter in *Speaking My Truth: Reflections on Reconciliation and Residential Schools*. Out of print. Digital copies can be downloaded from www.speakingmytruth.ca

"Rethinking Dialogue and History: The King's Promise and the 1906 Aboriginal Delegation to London," Keith Thor Carlson, *Native Studies Review* 16, no. 2 (2005).

Salish Weaving, Primitive and Modern, Oliver N. Wells, 1969, Oliver N. Wells, Sardis, BC.

The Amazing Mazie Baker: The Squamish Nation's Warrior Elder, Kay Johnston, 2016, Caitlin Press, Halfmoon Bay, BC.

The Earth's Blanket: Traditional Teachings for Sustainable Living, Nancy J. Turner, 2005, University of Washington Press, Seattle, WA.

The Face Pullers: Photographing Native Canadians 1871-1939, Brock Silversides, 1995, Fifth House, Saskatoon, SK.

The Indian Act 1876. www.aadnc-aandc.gc.ca/DAM/ DAM-INTER-HQ/STAGING/ texte-text/1876c18_ 1100100010253_eng.pdf

The Last Great Sea: A Voyage through the Human and Natural History of the North Pacific Ocean, Terry Glavin, 2000, Greystone Books, Vancouver, BC.

The Master and his Emissary: The Divided Brain and the Making of the Western World, Iain McGilchrist, 2010, Yale University Press, New Haven, CT.

The Rush For Spoils: The Company Province 1871-1933, Martin Robin, 1972, McClelland & Stewart Ltd., Toronto, ON.

The Sea Among Us: The Amazing Strait of Georgia, Richard Beamish and Gordon McFarlane, 2014, Harbour Publishing, Madeira Park, BC.

The Spirit of the Coast Salish, Sheila Thompson and Louise Steele, 1987, Creative Curriculum Incorporated, Vancouver, BC.

The "UN Declaration on the Rights of Indigenous Peoples." Adopted by the General Assembly on September 13, 2007 by a majority of 144 states in favour, 4 votes against (Australia, Canada, New Zealand, and the United States): www.un.org/esa/socdev/unpfii/documents/DRIPS_en.pdf

The Universe Story: From the Primordial Flaring Forth to the Ecozoic Era — A Celebration of the Unfolding of the Cosmos, Brian Swimme and Thomas Berry, 1992, Harper, San Francisco, CA.

Sound Investment: Measuring the Return on Howe Sound's Ecosystem Assets, Michelle Molnar, 2015, David Suzuki Foundation.

Squamish-English Dictionary, Squamish Nation Education Department, North Vancouver, 2011, University of Washington Press, Seattle, WA.

Squamish: The Shining Valley, Kevin McLane, 1999, Merlin Productions, Squamish, BC.

"The Squamish Indian Land Use and Occupancy" a study prepared by Randy Bouchard and Dorothy Kennedy, 1986, BC Indian Language Project, Victoria, BC.

"Telling Truths and Seeking Reconciliation," a chapter by Rupert Ross, in *Speaking My Truth: Reflections on Reconciliation and Residential Schools*, no longer in print but available for download at www.speakingmytruth.ca

We Were Born with the Songs Inside Us, Katherine Palmer Gordon, 2013, Harbour Publishing, Madeira Park, BC.

Xay Temixw (Sacred Land) Land Use Plan: www.squamish.net/about-us/our-land/xay-temixw-sacred-land-land-use-plan

Rich Duncan

ABOUT THE AUTHOR

Pauline Le Bel is an award-winning novelist, an Emmy-nomi-
nated screenwriter and the author of *Becoming Intimate with the
Earth*. A singer-songwriter, she has produced five CDs of her
original music. In an earlier life, she was called "a musical instru-
ment linked to a soul" for her passionate portrayal of chanteuse
Edith Piaf in a play she co-wrote. Today, she is the creative director of a reconciliation initiative, Knowing Our Place. She lives
on Bowen Island. For more information about Pauline, please
visit her website: www.paulinelebel.com

To engage in ongoing discussion and updates about Howe
Sound/Atl'kitsem, please visit: www.whaleinthedoor.com

This book is set in Monotype Bembo. Based on a design cut around 1495 by Francesco Griffo for Venetian printer Aldus Manutius, Bembo is named for Manutius's first publication in 1496 — a small book by poet and cleric Pietro Bembo.

The text was typeset by Vici Johnstone.

Caitlin Press, Fall 2017.